T0135216

Advances on Robotic Item Picking

Albert Causo • Joseph Durham • Kris Hauser
Kei Okada • Alberto Rodriguez
Editors

Advances on Robotic Item Picking

Applications in Warehousing & E-Commerce Fulfillment

Editors
Albert Causo
Robotics Research Centre
Nanyang Technological Univ
Singapore, Singapore

Joseph Durham
Amazon Robotics LLC
Amazon (United States)
North Reading, MA, USA

Kris Hauser
Duke University
Durham, NC, USA

Kei Okada
University of Tokyo
Tokyo, Japan

Alberto Rodriguez
Massachusetts Institute of Technology
Cambridge, MA, USA

ISBN 978-3-030-35681-1 ISBN 978-3-030-35679-8 (eBook)
https://doi.org/10.1007/978-3-030-35679-8

This Springer imprint is published by the registered company Springer Nature Switzerland AG.
The registered company address is: Gewerbestrasse 11, 6330 Cham, Switzerland

Preface

This book was borne to capture the ideas and solutions that were displayed and deployed at the Amazon Robotics Challenges. During the 3 years that the Challenge ran from 2015 to 2017, more than 50 unique teams have participated. Some have participated once or twice, while others have been there every year. These teams shared nothing in common at the beginning but by the end of each competition, and over the course of 3 years, participating robotics engineers and scientists started to form an informal community which share a common passion of solving the problem of robotic picking as applied to the e-commerce and warehouse fulfillment scenario.

What Amazon Robotics started in 2015 with the first Amazon Picking Challenge, as it was called then, evolved into a concerted effort to solve the challenges of robotic picking geared more for industry deployment rather than academic curiosity. The chapters in this book have been arranged by year to show the evolution of the solutions and the challenge specifications, itself a reflection of the ever-tightening requirement of robots when deployed in industrial applications.

Each chapter provides a glimpse to the philosophy and strategies behind the teams' choices of hardware components and software approach. Each team provides a description of their overall system design and discusses in more detail the identification and pose determination algorithms, the hardware and software strategies for grasping, the motion planning approach, and error recovery techniques.

We hope that this book provides exciting reading materials to those keen in robotic item picking problem. Papers from winning teams are included in this book, and reflections on both successes and failures shared by the authors themselves are a good starting point for anyone interested to work on the topic.

Lastly, we would like to acknowledge all the teams who have contributed to this book. We realize that it is not so easy to write about your own work a few years after the fact. We would also like to thank the teams who shared their experience and expertise during the workshops organized after each competition. And last but not least, our sincerest thanks to Amazon Robotics, which through its support of the competitions has pumped adrenaline and helped focus the robotics research community's attention to existing industry pain points when it comes to item picking.

Singapore, Singapore — Albert Causo
North Reading, MA, USA — Joseph Durham
Durham, NC, USA — Kris Hauser
Tokyo, Japan — Kei Okada
Cambridge, MA, USA — Alberto Rodriguez
October 2019

Contents

Team CVAP's Mobile Picking System at the Amazon Picking Challenge 2015

Kaiyu Hang, Francisco E. Viña B., Michele Colledanchise, Karl Pauwels, Alessandro Pieropan, and Danica Kragic

1 Introduction

This paper presents a comprehensive description of the framework developed by the team of the Robotics, Perception and Learning laboratory (RPL) for the competition organized by Amazon in Seattle in 2015. We assume that the reader is familiar with the competition and the set of rules of the edition in 2015. The paper is structured as follows: the mobile robotic platform used in the competition is described in Sect. 2. Section 3 will present the strategy we employed to maximize the performance given the limitation of the robot used. Behavior trees, our modular and reactive tool for planning, will be described in Sect. 4. Section 5 will present how the shelf is detected and how such information is used to determine the movement of our robotic platform within the working cell. Finally, we will explain how the objects are detected in Sect. 6 and grasped in Sect. 7.

2 Platform

We used the PR2 research robot from Willow Garage shown in Fig. 1 for the 2015 APC competition. The robot consists of two 7-DOF manipulators attached to a height-adjustable torso and equipped with parallel-jaw grippers. A pan-tilt

All authors contributed equally in this work.

K. Hang · F. E. Viña B. · M. Colledanchise · K. Pauwels · A. Pieropan (✉) · D. Kragic
KTH Royal Institute of Technology, Stockholm, Sweden
e-mail: kaiyuh@kth.se; fevb@kth.se; miccol@kth.se; kpauwels@kth.se; pieropan@kth.se; dani@kth.se

A. Causo et al. (eds.), *Advances on Robotic Item Picking*,
https://doi.org/10.1007/978-3-030-35679-8_1

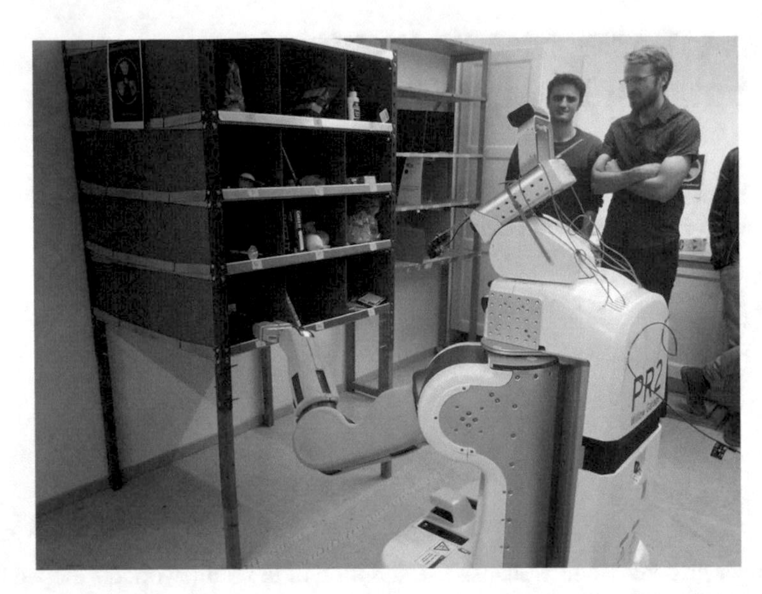

Fig. 1 Team CVAP's PR2 robot platform used for the Amazon Picking Challenge 2015 in Seattle, USA

controllable robot head is located on the upper part of the robot and equipped with a set of RGB cameras and a Kinect camera which we used in our object segmentation system. The robot is also equipped with a tilt-controllable laser scanner located right beneath the head which we also used for object segmentation purposes. A set of four wheels at the base provides the robot with omnidirectional mobility, which we exploited in order to control the robot back and forth from the shelf to pick up objects and release them in the target box.

We provided some minor hardware modifications to the PR2 robot in order to address some of the challenges of the picking task, namely custom-made extension fingers for the parallel gripper in order to be able to reach further inside of the bins of the shelf and a high resolution web camera which we attached on the PR2's head in order to get a richer set of image features for our Simtrack vision system to detect the target objects in the shelf.

The robot ran on an Ubuntu 12.04 computer with a real-time Linux kernel that provides 1 KHz manipulator control. All the high level task execution, perception, grasping, and manipulation software components were developed under the Robot Operating System (ROS).

3 Strategy

In 2015 the competition consisted in picking one object from each of the 12 bins on the shelf and place the requested items in a container located inside the define working area as shown in Fig. 2. Each bin could contain from one up to four objects

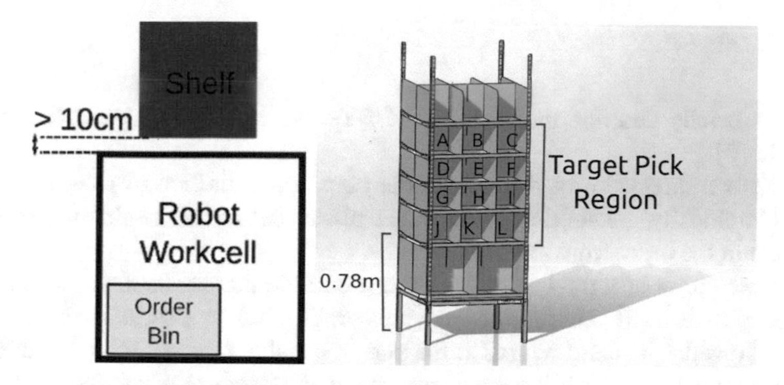

Fig. 2 *Left*: Work cell area of the competition. *Right*: The shelf layout

making the recognition and grasping of objects increasingly difficult. The design of our strategy took into consideration three main limitations of the robotic platform that was available at our laboratory. First, the PR2 could not reach the highest level of the shelf; therefore, we excluded all the objects located from bin A to C from the strategy (Fig. 2). Second, two of the competition's objects could not be picked at all since their volumes were bigger than the maximum aperture of the PR2 gripper. Third, raising the torso of the PR2 to reach the desired level of operation was a very expensive operation taking up to 30 s to raise the platform from its lowest position to its highest. Clearly, such action could have a huge impact given the time restriction of 20 min. Therefore, our strategy consisted in prioritizing the operation in each bin according to the level they were located on the shelf starting from the lowest (e.g., J, K, L). Our strategy also focused first in completing the tasks in bins with one or two objects leaving the most complex scenarios at the end of the task queue no matter what level in shelf they were located. Moreover, in order to limit even more the impact of moving the PR2's torso, such operation has been executed meanwhile the robot had to move from the shelf to the order bin after a successful grasp or vice versa after dropping an object in the order bin.

4 Behavior Trees

Behavior trees (BTs) are a graphical mathematical model for reactive fault tolerant action executions. They were introduced in the video-game industry to control non-player characters, and they are now an established tool appearing in textbooks [1, 2] and generic game-coding software such as Pygame, Craft AI, and the Unreal Engine. In robotics, BTs are appreciated for being highly modular, flexible, and reusable, and have also been shown to generalize other successful control architectures such as the subsumption architecture, decision trees, and the teleo-reactive paradigm [3].

4.1 Semantic

Here we briefly describe the semantic of BTs. An exhaustive description can be found in [3].

A BT is a directed rooted tree with the common definition of *parent* and *child* node. Graphically, the children of nodes are placed below it. The children nodes are executed in the order from left to right.

The execution of a BT begins from the root node that sends *ticks*[1] with a given frequency to its (only) child. When a parent sends a tick to a child, the execution of this is allowed. The child returns to the parent a status *running* if its execution has not finished yet, *success* if it has achieved its goal, or *failure* otherwise.

There are four types of internal nodes (fallback, sequence, parallel, and decorator) and two types of leaf nodes (action and condition). Below we describe the execution of the nodes used in our framework.

Fallback The fallback node sends ticks to its children from the left, returning success (running) when it finds a child that returns success (running). It returns failure only when all the children return failure. When a child returns running or success, the fallback node does not send ticks to the next child (if any). A fallback node is graphically represented by a box labeled with a "?", as in Fig. 3.

Sequence The sequence node sends ticks to its children from the left, returning failure (running) when it finds a child that returns failure (running). It returns success only when all the children return success. When a child returns running or failure, the sequence node does not send ticks to the next child (if any). A sequence node is graphically represented by a box labeled with a "→", as in Fig. 3.

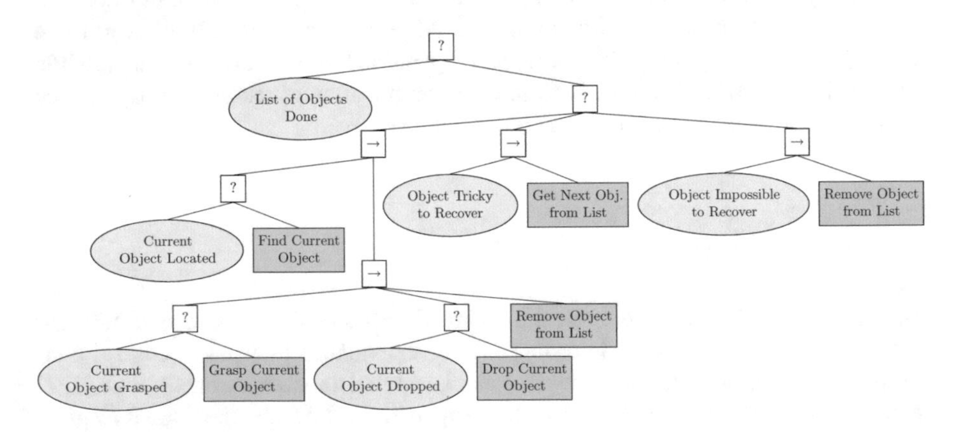

Fig. 3 BT defined at the design stage

[1]A tick is a signal that allows the execution of a node.

Action The action node performs an action. It returns running while the action is being performed. It returns success of the action is completed correctly, otherwise, it returns failure. An action node is graphically represented by a rectangle labeled with the name of the action, as in Fig. 3.

Condition The condition node checks if a condition is satisfied or not, returning success or failure accordingly. An action node is graphically represented by an ellipse labeled with the name of the condition, as in Fig. 3.

Internal Nodes with Memory There also exists a memory version of the internal nodes described above. Nodes with memory always tick the same child until this returns success or failure then it evaluates (if applicable, according to the node's logic) the next children. Such control flow nodes are graphically represented with the addition of the symbol "$*$" (e.g., a sequence node with memory is graphically represented by a box with a "$\rightarrow$*").

4.2 BTs in APC

In our framework, the use of BTs allowed us to have a control architecture that is:

- *Reactive:* The reactive execution allowed us to have a system that rapidly reacts to unexpected changes. For example, if an object slips out of the robot's gripper, the robot will automatically stop and pick it up again without the need to re-plan or change the BT; or if the position of an object is lost, the robot will re-execute the routine of the object detection.
- *Modular:* The modular design allowed us to subdivide the behavior into smaller modules that were independently created and then used in different stages of the project. This design allowed our heterogeneity developers' expertise, letting developers implement their code in their preferred programming language and style.
- *Fault Tolerant:* The fault tolerant allowed us to handle actions failure by composing fallback plan whenever an object was not correctly grasped or dropped.

We initially designed a BT that fully exploits the aforementioned advantages, then we made some simplification due to time and platform constraints, as described later. Figure 3 shows the BT defined at the design stage. The BT reads the value of *current object* to pick from the ROS parameter master. An external ROS node that implements our strategy (see Sect. 3), called *strategy* node, provides such value. The BT first checks if there is an object left to pick. In such case, the BT executes the following behavior. It tries to detect, grasp, and drop the object into the bin. If it succeeds, it requests to the strategy node to remove the object from the list of objects to grasp. If it fails, it assesses whether the object is easy, tricky, or impossible to recover. If the object is easy to recover, the aforementioned behavior is repeated; if tricky, the BT requests the strategy node to move on to the next object to pick. If the object is impossible to recover, the BT requests the strategy node to remove the object from the list of objects to grasp. Note that the BT in Fig. 3 implements intrinsically a reactive behavior. While the robot is grasping or dropping the object, it continually checks if the object position is still known. If, for example, while grasping, the object position becomes unknown (e.g., the open grippers push the object away), the BT will abort the grasping or dropping action to return to detect the object.

As mentioned above, we made some modification to the original BT. In particular, the low-level implementation for some condition nodes was difficult to achieve. For example, it was hard to discriminate if the object was correctly dropped or whether the object is tricky or impossible to recover. Moreover, when the object was not dropped correctly, most of the time was on the floor, hence impossible to recover due to the limited capability of the robot. Hence, after some iteration during the development phase, the BT had some simplifications. Figure 4 shows the actual BT used at the APC. Since we chose to not implement some condition nodes, we could not achieve a fully reactive behavior; hence, we chose to use control flow nodes with memory (see Sect. 4.1).

We added two intermediate actions: *Torso Positioning* that adjusts the height of the robot according to the bin where the object is positioned, and *Pregrasp* that executes a predefined arm movement to ease the grasp search. Moreover, if the grasping fails, we let the robot go to the next object on the list without removing the current object, which could be tried after tying the other objects. We let the action drop always return success, hence after any attempt, the object will be removed from the list. Even if the drop does not succeed, the object would not be recoverable and the object will be anyway removed from the list.

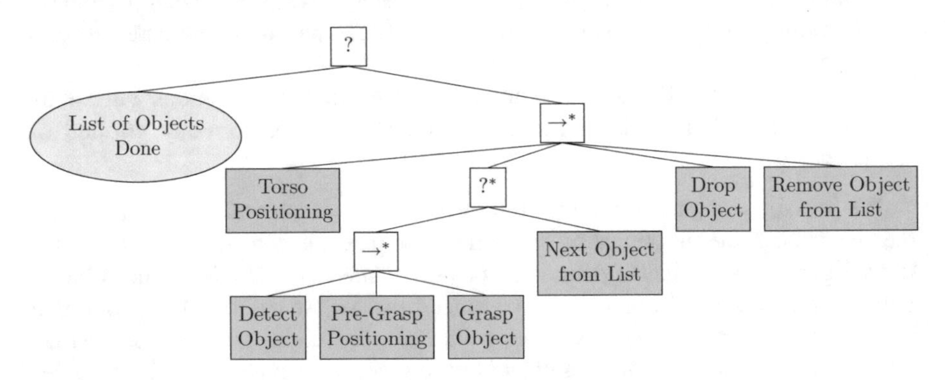

Fig. 4 Revised BT used at the APC

5 Shelf Localization and Base Movement

As described in Sect. 2, we used a PR2 as our robot platform. Since the arms' reachability of PR2 is relatively small in comparison to the shelf size, it is not feasible to define a fixed location for the robot to achieve the required task. As such, we have to exploit the mobility to enlarge the workspace, so that the robot would be able to reach and grasp from most of the shelf bins, as well as loading the grasped objects into the order bin.

For this, shelf localization, which serves as the only landmark in the work cell, is essential for our system to guide the robot navigating between different grasping positions. Since the robot movement accumulates localization errors, it is necessary to localize the shelf in real time to close the control loop for base movement.

As shown in Fig. 5, we use the base laser scanner to localize the two front legs of the shelf. Observing that the shelf is located in front of the robot, there are no other objects close by. Therefore, it is reasonable to find the closest point cluster and consider it as one of the legs, while the remaining cluster is considered as another leg. However, this is not a reliable procedure as there could be noise or other unexpected objects, e.g., human legs. As such, our shelf localization consists of two procedures as follows:

- *Detection:* Once the front legs are detected, before the system starts to autonomously work on assigned tasks, a human supervisor needs to confirm to the robot that the detection is correct. In case when the detection is incorrect, we need to clear the occlusions in front of the shelf until a confirmation is made.
- *Tracking:* While the robot is moving, given that we know the motion model of the robot, we update the shelf localization using a Kalman filter.

Fig. 5 *Left*: An example of shelf localization shown in rviz. The x and y coordinates of shelf_frame is defined as the center of two front legs, while the height of it is the same as the base_laser_link. *Right*: The bin frames are located at the right bottom corner of each bin

Having localized the shelf, we further estimate the shelf bin frames based on the known mesh model, see Fig. 5. As will be described in Sect. 7, knowing the bin frames will facilitate the grasp planning in our system.

6 Vision

6.1 *Simtrack*

We created high-quality 3D textured models from all objects provided for the challenge using the approach described in [4]. To summarize briefly, each object is placed on a textured background consisting of a collection of newspapers and approximately 40 pictures are taken from various angles. The object is then constructed using Autodesk123D catch services [5] and postprocessed to separate it from the background, remove holes, reduce the mesh resolution, etc. For the texture-based detection discussed next, we also extracted SIFT-keypoints [6] from images synthesized by rendering the model from multiple viewpoints. The final object models were used for detection and grasp planning.

A large proportion of the challenge objects contained sufficient texture for keypoint-based pose detection. When such objects were present in the bin, we relied on a texture-based pose estimation method. We used high resolution images obtained from a Logitech C920 webcam mounted on the robot head. We processed a central 1152×768 pixel region of the full 2304×1536 input image for object detection. The robot head was always oriented towards the bin under consideration so that this region of interest was guaranteed to contain the objects. Our object pose detection framework, SimTrack, first extracts SIFT features from the image and matches these to the database of SIFT features extracted from the object models in the modeling stage. In the detection stage, the objects are identified, and a rough initial pose is obtained. This pose is then iteratively refined by rendering the textured object models at the current pose estimate and locally matching their appearance to the observed image [7]. SimTrack uses a phase-based optical flow algorithm in this refinement (or tracking) stage to reduce sensitivity to intensity differences between model and observed image. SimTrack exploits the rich appearance and shape information of the models and correctly handles occlusions within and between objects. Some example detections are shown in Fig. 6. SimTrack is publicly available as a ROS module [8].

6.2 *Volumetric Reasoning*

We designed a fallback strategy based on 3D point cloud volumetric reasoning in case our system could not recognize the objects using SimTrack. Knowing the

Fig. 6 Model-based object detection in action. The rendered models are overlaid in green

location of the shelf and the bins, as shown in Fig. 5, we use the PR2 tilting laser scanner mounted on the torso to build an accurate point cloud within the target bin. We use the 3D model of the shelf and its estimated location to remove any point belonging to the shelf from the constructed point cloud to obtain a cloud of the inside of a bin. Then, we apply Euclidean clustering to generate plausible object hypothesis. Given that the number of objects in the bin is known, we apply our clustering strategy iteratively, increasing the distance threshold, in order to obtain as many clusters as the expected number of objects. Once the clusters are found we formulate the problem of finding the right object to pick as a bipartite graph matching problem where the nodes on one side are the found clusters characterized by their volumes and the nodes on the other side are the 3D models of the expected objects and their corresponding volumes.

7 Picking and Placing

For robot arm motion planning we used MoveIt! and its default randomized motion planners. Even though MoveIt! provides seamless integration with ROS-enabled robots, we faced some difficulties when applying it in our PR2 APC setup. The sampling based planners (e.g., RRTConnect) often had difficulties in finding collision free trajectories to move the robot arm in constrained spaces such as the interior of the bins. Even when it did find suitable trajectories, they would often generate very large joint displacements even for small end-effector Cartesian motions, which increased the chance of generating collisions with the shelf. We partially alleviated these problems by exploiting the simpler 2D kinematic motion control of the robot base, e.g., to reach and retreat from the shelf. Trajectory optimization based motion planners or a more specialized kinematic structure of the robot are possible solutions for addressing these issues.

Once the target object is detected, the object frame is obtained from our vision system. Given that the PR2 is equipped with parallel grippers, we must then decide from which position and in which direction the gripper is going to approach the object in order to grasp the object in a feasible way. For this, we always project the object frame based on the bin frame such that the X axis of the object frame points inwards the shelf, and the Z axis points downwards, as illustrated in Fig. 7. In order to ensure that the object can be approached without generating collisions, we manually labeled all objects offline so that the X axis returned from the vision system can always be approached with the gripper, i.e., the gripper opening is wide enough to grasp the corresponding side of the object. Thereafter, similarly to the grasping stages in [9], our grasping system works in the following steps:

- *Pre-grasping:* The gripper is posed in front of the bin and aligned in X axis of object frame; the gripper's opening direction is aligned in the XY plane of object frame.
- *Approaching:* The gripper approaches the object in X direction until a pre-determined depth in the object frame is reached.
- *Grasping and lifting:* The gripper closes and lifts the object to a pre-determined height. At this point, we read the gripper's joint angle to check whether the object is grasped. If not, the system will return failure and give control back to the behavior tree. Otherwise, it returns success and continue to the next step.
- *Post-grasping:* When the object is grasped, it is not trivial to plan a motion to move the arm out of the shelf, especially for large objects. Therefore, we first move the object to the bin center, and then move the whole robot backwards. Once the object is out of the shelf, the robot goes to the order bin and loads the object into it.

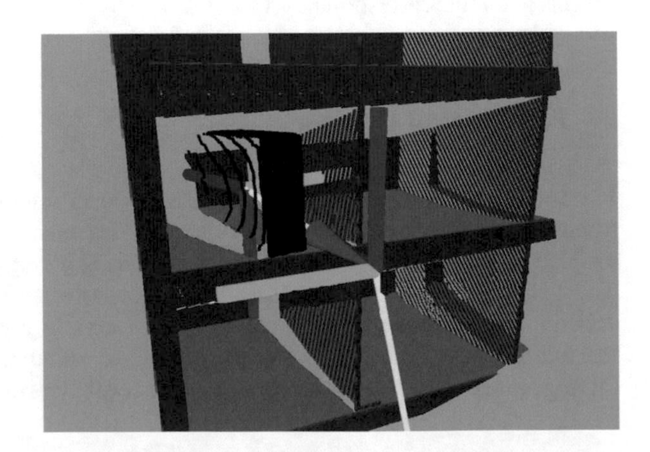

Fig. 7 Independent of the object's pose, the object's frame within the shelf bin is always projected in terms of the bin's frame, so that the X axis of the object frame points inwards the shelf, and the Z axis points downwards

In cases when the target object is very small, e.g., tennis ball and duck toy etc., the approaching procedure shown above is not able to reach the object due to the collision with the bin bottom. In those cases, our system will first try to approach the gripper to the area above the object, and then move downwards to reach the desired grasping pose. It is worthwhile noticing that dividing the grasping motion into steps significantly increased the success rate for MoveIt! to find solutions within a short time limitation, since the above steps effectively guides the motion planner through narrow passages constrained by the small bin areas.

8 Conclusion

We presented the framework used in the APC 2015 and the challenges that we faced. Overall, perception was the most challenging component for the competition although the challenges inherent to manipulation and grasping should not be underestimated. We can propose the following suggestions to researchers and engineers designing bin-picking robotic systems given our experience at the APC 2015:

- *Robot kinematic design.* In order to facilitate motion control of the robot manipulator and minimize the risk of generating collisions it is best to design its kinematic structure to be completely tailored to the task instead of a general purpose humanoid kinematic configuration such as our PR2 robot.
- *Grasping via suction devices.* Using suction devices for grasping would have greatly increased our grasp success rate. Suction is the de facto solution to robot grasping in many industrial applications, including robot bin picking. Part of the reason for this is that suction is in general more tolerant to errors in perception and motion control since it is less demanding in terms of where to position the gripper relative to the object to be grasped. An object whose size is comparable to the maximum opening of a robot's parallel gripper requires very precise positioning of the gripper, otherwise the robot will unintentionally push the object away when approaching it. On other hand, suction devices have a higher likelihood of yielding a successful grasp under positioning errors as long as it is still in contact with part of the object's surface. This also means that the requirements for the perception system could be lowered, e.g., by segmenting identified flat object surfaces instead of performing full 6DOF pose estimation and requiring less accurate object pose estimation.
- *Deep learning for object detection.* Given the impressive results of convolutional neural networks in recent years, our perception system would greatly improve its recognition accuracy using a network trained to recognize the objects used in the challenge. There are well-known networks that can localize objects in real time. It would be interesting to see how deep learning can be leveraged to estimate the full 6-DOF pose of the objects.

Acknowledgements This work has been supported by the Swedish Research Council (VR), the Swedish Foundation for Strategic Research (SSF), and the EU FP7 program. The authors gratefully acknowledge the support.

References

1. Millington, I., Funge, J.: Artificial Intelligence for Games. CRC Press, New York (2009)
2. Rabin, S.: Game AI Pro. CRC Press, Boca Raton (2014), ch. 6. The Behavior Tree Starter Kit
3. Colledanchise, M., Ögren, P.: Behavior Trees in Robotics and AI: An Introduction. Chapman and Hall/CRC Artificial Intelligence and Robotics Series. CRC Press, Taylor & Francis Group, Boca Raton (2018)
4. Pokorny, F.T., Bekiroglu, Y., Pauwels, K., Butepage, J., Scherer, C., Kragic, D.: A database for reproducible manipulation research: CapriDB – Capture, Print, Innovate. Data Brief **11**, 491–498 (2017)
5. Autodesk. http://www.123dapp.com. Accessed 01 May 2017
6. Lowe, D.G.: Distinctive image features from scale-invariant keypoints. IJCV **60**(2), 91–110 (2004)
7. Pauwels, K., Kragic, D.: SimTrack: a simulation-based framework for scalable real-time object pose detection and tracking. In: IEEE/RSJ International Conference on Intelligent Robots and Systems, Hamburg, Germany (2015)
8. SimTrack. https://github.com/karlpauwels/simtrack. Accessed 01 May 2017
9. Hang, K., Li, M., Stork, J.A., Bekiroglu, Y., Pokorny, F.T., Billard, A., Kragic, D.: Hierarchical fingertip space: a unified framework for grasp planning and in-hand grasp adaptation. IEEE Trans. Robot. **32**(4), 960–972 (2016)

Team UAlberta: Amazon Picking Challenge Lessons

Camilo Perez Quintero, Oscar Ramirez, Andy Hess, Rong Feng, Masood Dehghan, Hong Zhang, and Martin Jagersand

1 Introduction

The Amazon challenge involves solving a real-world pick-and-place problem with the state-of-the-art hardware, software, and algorithms. Furthermore, the challenge forced robotics researchers to provide a portable solution that should work reliably in a different place from their lab environment.

The picking challenge required teams to pick out specific items from a shelf and to place those items into a container. While the objects that would be present for the challenge were known beforehand, the specific objects to be grasped at the competition were given at the start of the marked trial (Fig. 1). To complete the task, the autonomous robotic system must fulfill a work order that specifies objects to grasp from a specific bin. To do so the system has to identify and locate the requested objects from the bin. Then, execute autonomously a planning and manipulation strategy to grasp the objects and put it in a container without disturbing the environment (e.g., hitting the shelf, damaging other objects inside the bin, picking up wrong objects). There were 26 international teams in the competition with diverse approaches. In our approach we developed our solution based on the available robot hardware (7DOF WAM arm with Barrett hand) in our research lab. We identified the following challenges:

- Although our WAM arm has a large work-space (spherical work-space of approximately 2 m in diameter), it could not reach the inside of each bin on the shelf, thus it was only capable of grasping objects in the front face of the shelf.

C. P. Quintero (✉) · O. Ramirez · A. Hess · R. Feng · M. Dehghan · H. Zhang · M. Jagersand
Department of Computing Science, University of Alberta, Edmonton, AB, Canada
e-mail: caperez@ualberta.ca

A. Causo et al. (eds.), *Advances on Robotic Item Picking*,
https://doi.org/10.1007/978-3-030-35679-8_2

Fig. 1 The UAlberta team during their competition trial

- In the few bins that the arm was able to reach inside, the Barrett hand was too big to be actuated while inside the bin, this was commonly identified by many teams [1, 2].

To address these problems, we thought of different solutions such as adding a mobile base to the robot or adding an extension to the robot arm to reach inside the bins.

After receiving the application equipment (Amazon shelf and objects) we tested the robot reaching capacity, and we noticed that our robot is able to reach the external face of all bins, but the robot arm is not long enough to reach inside them. The bin dimensions are $21 \times 28 \times 43$ cm. Knowing that we are not using a mobile base, we had to either increase the reachability of our arm or guarantee that the target object is always going to be in the graspable reaching region of the arm, i.e., close to the border of the bin face and centered.

Through our design process we quickly prototyped and iterated on our approach. We began using a binocular RGB camera setup and applying uncalibrated image based visual servoing (IBVS) for the robot control [3]; the precision obtained with this method was high compared to RGB-D methods; however, the control algorithm relied heavily on object tracking, making the system unstable and unreliable [4]. Then, we decided to include an RGB-D sensor and implement object localization by clustering based on Euclidean distance. Unfortunately, the bottom part of the shelf has a shiny surface that the depth sensor failed to sense correctly. In our final competition system our robot was instrumented with two eye-in-hand cameras,

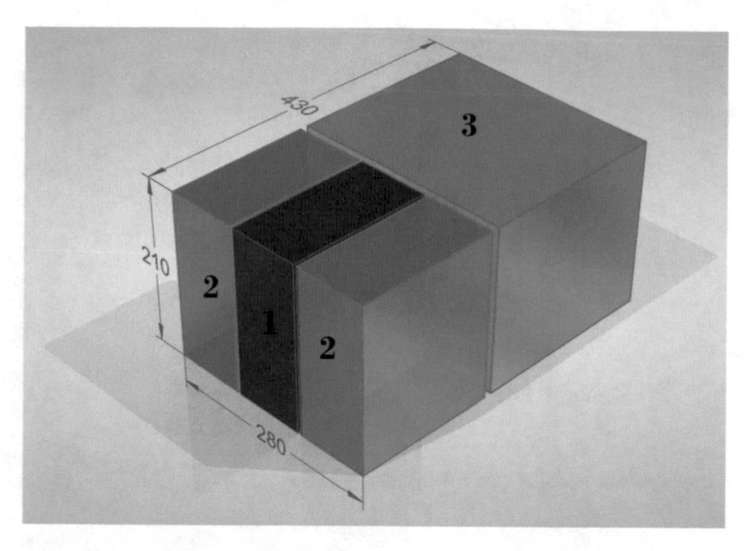

Fig. 2 Bin volume divided in regions of operation. (1) Graspable region. (2) Side regions. (3) Back region

a LiDAR, and a custom-made object push–pull mechanism. Based on the initial challenge description, our first experimental iterations, and our available hardware and resources we designed the following strategy.

2 Strategy

This section describes our grasping strategy. Due to physical limitation of our arm/gripper and also to simplify our grasping strategy, we divide each bin into three regions: (1) graspable region, (2) side regions, and (3) back region as shown in Fig. 2. Our gripper is only able to pick an object if it is in region (1). Assuming that the bin number and the object to pick from that bin are given, the goal is for the robot to go to the bin, locate the object, and if it is in the back region or side regions, bring it to the center region first and then grasp it and finally place it inside a container.

2.1 Pulling–Pushing Mechanism

To provide our robot with the capacity of reaching inside the bin and relocate the target object into the grasping region, we designed a linear actuator with two design requirements. (1) It has to be able to reach the end of the bins from the outside of the bin face, i.e., have the capacity to elongate and retract approx. 45 cm and

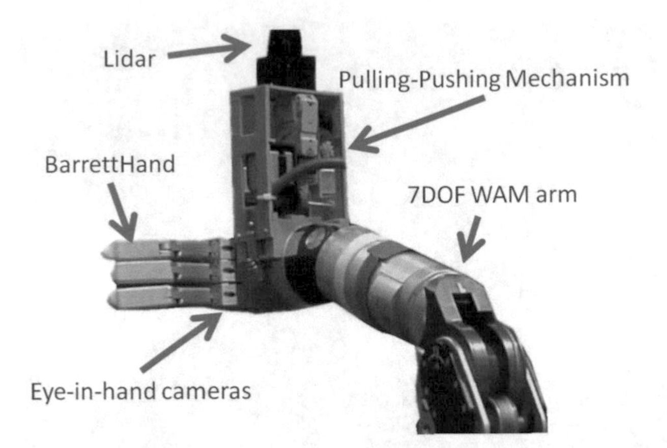

Fig. 3 Overview of our instrumented robot hand and arm

when it is retracted not occupy more than 9 cm (this is the space that can be used on top of the robot wrist without affecting the robot movement); (2) Capability of supporting collisions without damaging the robot arm or the environment. To fulfill these requirements we designed the pulling–pushing mechanism shown in Figs. 3 and 7. These type of mechanisms have been used before in early satellite technology, due to frequency limitation antennas of 45 m where required, but before deployment these antennas had to fit into a 1 m diameter satellite. We designed a simple roller based mechanism that pulls a metallic measuring tape that is used as a linear actuator. With this mechanism we have the capacity of pulling, pushing, and relocating objects from regions two and three to region one (see Fig. 2). During our testing we also determined that we were able to push and pull weights up to about 1 kg which was well within our requirements.

3 System Description

A state diagram of our system is shown in Fig. 4. In state 1 the system reads in a work order from a provided JSON file and prioritizes the picking order of the bins based on the particular object difficulty and our confidence of a grasp success given the object distribution and bin placement.

Then, in state 2 the robot moves to the target bin and runs the detection algorithm. For this algorithm we pre-processed all objects for a total of 15,000 training images generated by applying multiple transformations to augment the size of our dataset. This is used to train 25 support vector machine classifiers (SVMs), used in a 1 vs rest training configuration. Our algorithm uses a sliding window and selects the point where the confidence of the target object is highest using the SVMs. An example is shown in Fig. 5, where the selected object is the "kong_duck_dog_toy" (Fig. 5c).

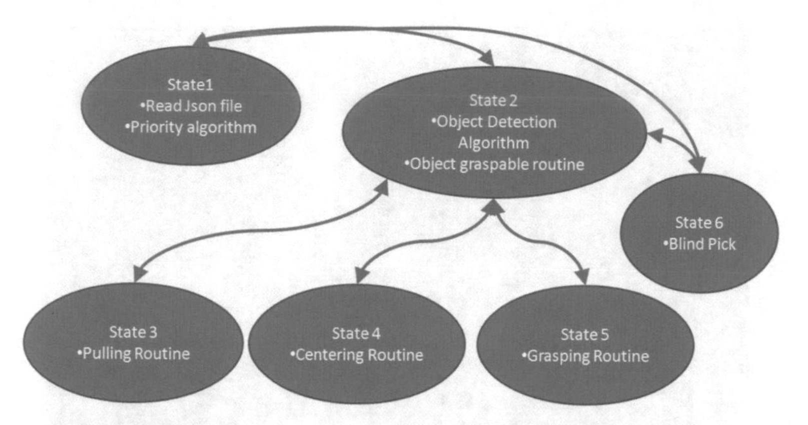

Fig. 4 State diagram

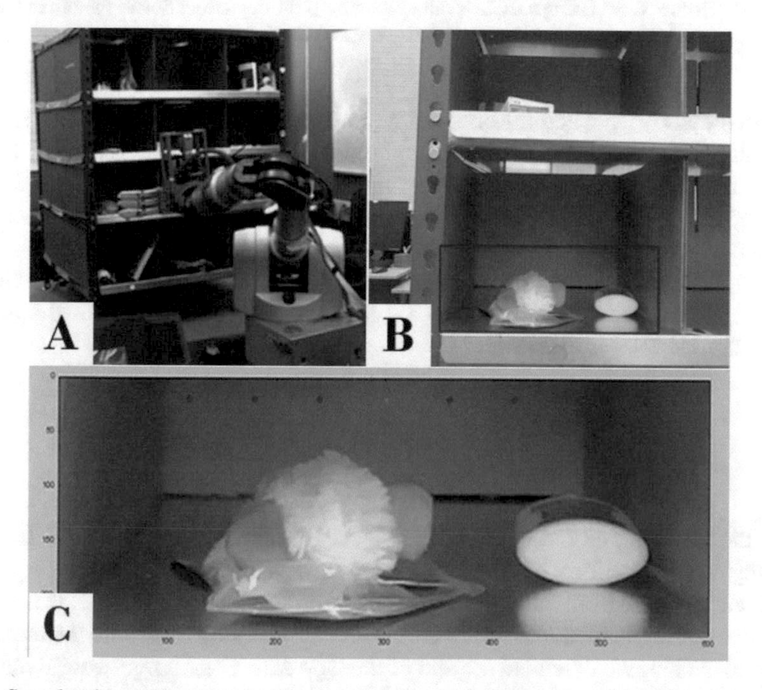

Fig. 5 Sample object detection. In this case the system (**a**) attempts to identify and locate the yellow duck (**b**) as marked by the green color dot in (**c**) a grasping location is found from the sliding window

The end-effector of the arm is moved in a way that the camera (located at the palm of the gripper) is facing the desired bin (Fig. 5b). Then, the object detection algorithm identifies the object and its location in the 2D image. The green dot in Fig. 5c shows position of the sliding window at which the confidence of having detected the duck is the highest. The output of the object detection algorithm is the location of the

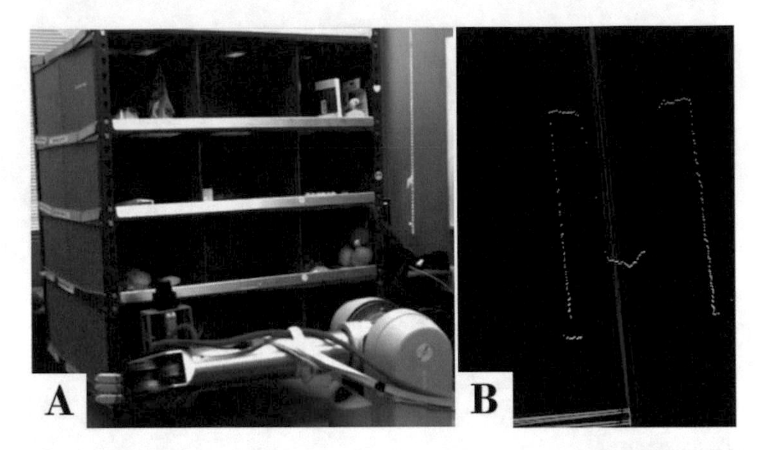

Fig. 6 Depth check. (**a**) Using a LiDAR our system finds the depth location in the bin of the target object. (**b**) Top-down view of the points seen by the LiDAR

duck in the 2D image, and the depth of the object is not known. To this end, we run the depth check routine, which is composed of a sequence of motions of the arm so that the 2D-LiDAR sensor is facing the target object (duck) in the desired bin (see Fig. 6a). The depth of the object is computed based on the output of the LiDAR scanner (Fig. 6b shows an overhead view of the occlusions within the bin).

Next, depending on the current region occupancy of the object, state 3 (pulling routine), 4 (centering routine), or 5 (grasping routine) is reach. Below is a description of each of these states.

3.1 Pulling Routine

Once the location and the depth of the object are obtained, we check if the object is in region (3), a pulling routine is activated in which our in-house device pulls the object closer to the front face of the bin. The pull/push mechanism is shown in Fig. 3 and uses a spring loaded measuring tape with a servo motor to extend the measuring tape into the bin. We extended the end of the tape with a small lightweight plate that we use to hook behind objects when performing the pull.

When the pulling routine is activated, the robot end-effector is located in a way that the measuring tape is above the object. We then extend the tape into the bin, lower the arm, and retract the tape. Given that the overall dimensions of the objects was known we performed the pulling action in a safe way to avoid dropping the objects. Figure 7 shows the robot performing this pulling action.

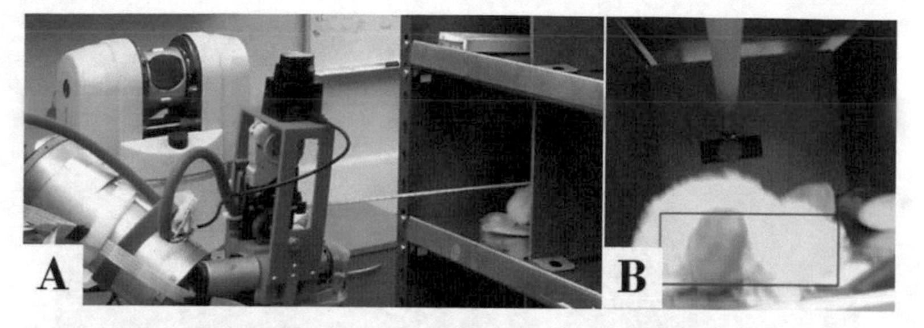

Fig. 7 Illustration of pulling routine. (**a**) the system extend the linear actuator to pull the yellow duck from region 3 to region 1. (**b**) Eye-in-hand view during pulling execution

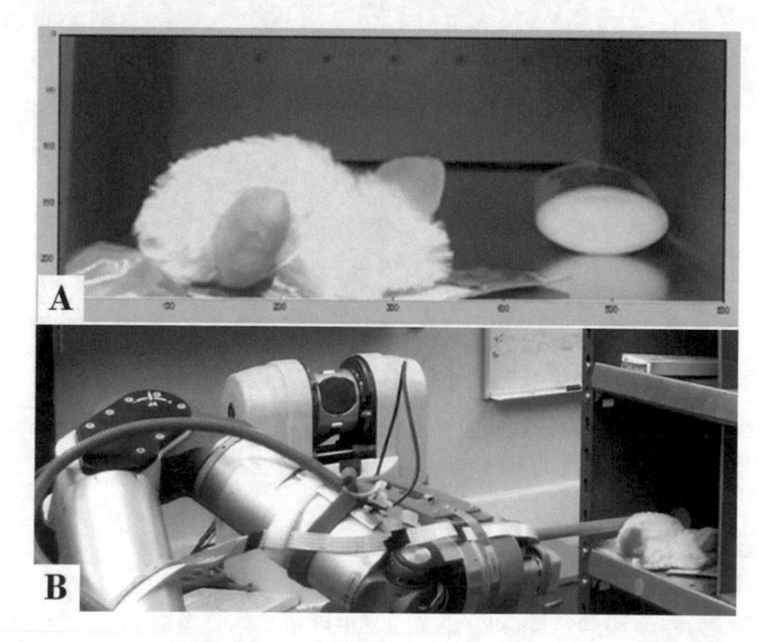

Fig. 8 Illustration of centering routine. (**a**) The duck location is estimated. (**b**) Robot uses the measuring tape to center the duck

3.2 Centering Routine

Due to the irregularity of the objects and also complex nature of the pulling mechanism, it is possible that the object ends-up in one of the side regions. Therefore, we check the location of the object again and in the case that it is not centered, we will activate an object centering routine, where the measuring tape is extended on the edge of the bin. We then use it as a lever to shift the object to the center of the bin. Figure 8 illustrates this routine.

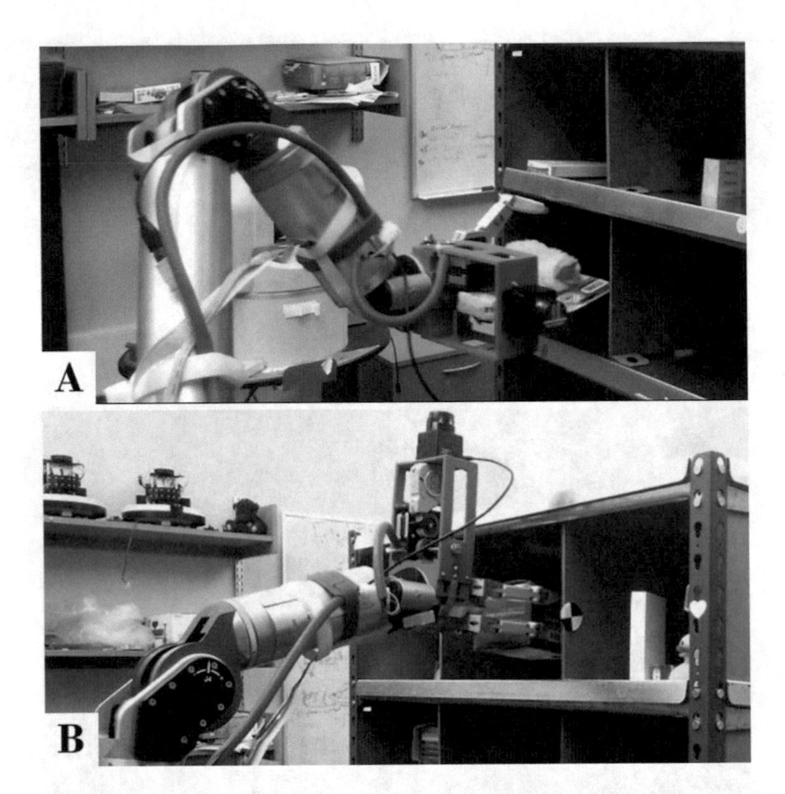

Fig. 9 Depending on the object size and mechanical properties, two different grasp modes can be used. (**a**) Pinch grasp. (**b**) Full finger grasp

3.3 Grasping Routine

Once the object is in the central region, it is within the reach of the gripper and the robot can grasp it. For the grasping, we use: (1) pinching mode (with two fingers) and (2) three finger grasping mode. The second mode is used for box-shaped/heavier items. These two modes are shown in Fig. 9.

Once grasped the end-effector was moved up and away from the bin to prevent grasps instability by hitting the edges of the bin. Once the object was fully outside the shelves bin we placed it in the target bin. A complete execution of our system can be seen in [5].

4 Conclusions

While working on this challenge we were able to take our robotics experience and applied it to a real-world problem. We developed most of our system during our personal time and with a budget of $500 dollars and access to our robot system and sensors.

The main lesson we gained was that in real life scenarios the best way to validate results is through experimental iterations. It is easy to overthink and stipulate about what will work, but when it comes to putting things into practice many unforeseen challenges arise. Iterating quickly and breaking things along the way (within reason) allowed us to quickly identify what the failure modes of our system were and gave us ideas as to how we should address them. Design, integration, and deployment of a practical robotic system like the one used for the amazon picking challenge are hard. Making the system reliable is even harder. This is something that roboticists have to consider more seriously if we want to see the robots in human environments.

References

1. Eppner, C., Höfer, S., Jonschkowski, R., Martın-Martın, R., Sieverling, A., Wall, V., Brock, O.: Lessons from the amazon picking challenge: four aspects of building robotic systems. In: Proceedings of Robotics: Science and Systems, AnnArbor, Michigan (2016)
2. Correll, N., Bekris, K.E., Berenson, D., Brock, O., Causo, A., Hauser, K., Okada, K., Rodriguez, A., Romano, J.M., Wurman, P.R.: Lessons from the amazon picking challenge (2016). Preprint. arXiv: 1601.05484
3. Jagersand, M., Fuentes, O., Nelson, R.: Experimental evaluation of uncalibrated visual servoing for precision manipulation. In: 1997 IEEE International Conference on Robotics and Automation, 1997. Proceedings, vol. 4, pp. 2874–2880. IEEE, Piscataway (1997)
4. Fomena, R.T., Quintero, C.P., Gridseth, M., Jagersand, M.: Towards practical visual servoing in robotics. In: 2013 International Conference on Computer and Robot Vision (CRV), pp. 303–310. IEEE, Piscataway (2013)
5. https://www.youtube.com/watch?v=SQJ5U3f3nWU&feature=youtu.be, May (2017)

A Soft Robotics Approach to Autonomous Warehouse Picking

Tommaso Pardi, Mattia Poggiani, Emanuele Luberto, Alessandro Raugi, Manolo Garabini, Riccardo Persichini, Manuel Giuseppe Catalano, Giorgio Grioli, Manuel Bonilla, and Antonio Bicchi

1 Introduction

More and more, industry entrusts robots for performing repetitive tasks, like heavy lifting, management of dangerous items, welding, and pick and place of objects with accurately known shape and position. However, some processes are still open problems for the automation and robotics research community and the only option today is to employ human operators. Picking operations are a prominent example of such kind of tasks which have not been automatized yet. These present several challenges like the wide variability of object's size and shape, unknown object pose, uncertain knowledge of the mechanical and geometrical properties of the object, and, further, the operation has to be carried out in an unstructured environment [2].

Nonetheless, a full automation of picking operation can have a huge impact on the logistics industry, improving operator's safety, and the reliability of the tasks. In order to push researchers' effort to address the automation of picking and place operations, in 2014 amazon.com called for the first edition of the Amazon Picking Challenge (APC), which took place in Seattle (WA) at the International Conference on Robotics and Automation, ICRA15.

T. Pardi (✉) · E. Luberto · A. Raugi
University of Pisa, Pisa, Italy

M. Poggiani
Soft Robotics for Human Cooperation and Rehabilitation, Istituto Italiano di Tecnologia, Genoa, Italy

M. Garabini · R. Persichini · M. Bonilla
Enrico Piaggio Research Center, University of Pisa, Pisa, Italy

M. G. Catalano · G. Grioli · A. Bicchi
Department of Advance Robotics, Istituto Italiano di Tecnologia, Genoa, Italy

A. Causo et al. (eds.), *Advances on Robotic Item Picking*,
https://doi.org/10.1007/978-3-030-35679-8_3

In the APC, a set of 24 household products, Fig. 5, were distributed among 12 bins; for each bin, a single target item had to be picked and placed in a storage box [1]. The committee assigned a score for each item picked, depending on: (1) how many other items were in the bin, (2) a complexity coefficient, which takes care of the difficulty of the grasp, (3) item damages during the operation. During the challenge, every team had 20 min for picking as many of the 12 target items as possible.

In this work, we describe the hardware and the control framework of the robot that has been developed to participate at the APC. The key element of our robotic solution is the employment of soft robotic devices that embed elastic elements in their design. Two main branches exist in soft robotics research. The first takes its inspiration largely from invertebrates and builds robots with continuously flexible materials. On the contrary, the other branch is inspired more by the vertebrate musculoskeletal system, and it aims to build robots in which compliance is concentrated mostly in the robot's joints. In literature, these robots are referred to as articulated soft robots or flexible-joint robots [16].

The robot proposed for the challenge (see Fig. 1) has two main sources of softness, a wrist composed of soft joints and an adaptable end-effector. The wrist is composed of three Qbmoves [5], which are Variable Stiffness Actuators (VSA) [18] that allow continuous changes of the motor's stiffness. As for the end-effector, we employed the Pisa/IIT SoftHand [6].

Fig. 1 Robot designed for competing in the APC challenge

These devices provide the robot with the capability of operating safely in an unstructured and constrained environment, and the ability to grasp objects with different sizes and shapes using the same end-effector.

A three degrees of freedom (DoFs) Cartesian robot has been developed for extending the operational space of the VSA wrist. On top of the wrist, a 3D camera is attached. The point cloud sensed by the camera is processed in order to obtain the pose of the object; then, a motion planner receives this information and generates a sequence of actions to accomplish the grasp.

2 Robot Design

The main feature demanded from the robot to develop for the challenge is the capability of interacting with the environment. Hence, a combination of Qbmoves and the Pisa/IIT SoftHand would have been very suitable to ensure the necessary elasticity for the task. However, due to torque limitations, designing a robot fully composed of soft components was impracticable. We, therefore, proposed a robot composed of two main subsystems: a wrist powered by three revolute Qbmoves, Fig. 2, and a Cartesian three DoF robot. As the end-effector, the robot mounts a Pisa/IIT SoftHand at the end of the soft wrist. This trade-off between soft and hard components allows the robot to interact with the environment keeping elasticity properties, and also overcome effort limitation issues.

In order to limit the maximum torque exerted on the actuators, a particular kinematic configuration has been chosen. Thanks to the serial configuration Yaw, Roll, and Pitch depicted in Fig. 2, in the rest configuration shown in figure, only the actuator responsible for pitching is under strain.

The Cartesian robot is powered by a Maxon DC 24 V motor for each DoF. Moreover, a physical gravity compensation mechanism along the vertical axis cancels the effort of the vertical motor in the resting condition.

In order to improve the grasp effectiveness to bigger objects, and extend the gripper grasping range, we equipped the robot with an electric piston, which is mounted right under the VSAs chain. On the tip of the piston, a U-shaped flat

Fig. 2 Soft wrist designed for the robot. It is composed of a chain of three VSA and the Pisa/IIT SoftHand

upholstered with neoprene is mounted. Exploiting this artificial constraint, the robot is able to hold objects during grasping and reach more robust grasp than only using the Pisa/IIT SoftHand.

Perceiving the environment is another essential point for the robot; information retrieved are used for navigation, planning, and objects recognition.

The robot features an Asus XtionPro as 3D vision stereo camera, depicted in Fig. 4c; it provides a point cloud together with RGB information.

We decided to mount the camera directly on the robot wrist to limit potential vision occlusion.

2.1 Qbmoves

The Qbmoves [5] are variable stiffness actuators (VSA) [18] designed to be modular, Fig. 3a. They have the capability of moving their output shaft while adapting the mechanical stiffness of the shaft itself independently, similar to natural musculoskeletal systems. The antagonistic behavior is obtained via two motors connected to the output shaft by a non-linear elastic transmission, which are realized with linear springs.

The Qbmoves have a cubic shape, and a Qbmove can be attached to another by an interconnection flange. Ideally, this feature allows to build robots with wide range of configurations, like Cartesian robot, spherical wrists, or articulated arms [8].

2.2 Pisa/IIT SoftHand

Soft end-effectors are grippers designed to be able to resist disarticulations and impacts. This ability allows this category of end-effectors to perform safe and effective interactions with the environment, like picking or manipulating objects, even in constrained environments.

To achieve such performance, the Pisa/IIT SoftHand [6] (Fig. 3b) is composed of small rolling components (fingers or phalanges) interconnected via elastic bands. These give to each articulation an intrinsic elasticity. The well-known concept of postural synergies [17] describes a set frequent motion primitive for grasping operations. The Pisa/IIT SoftHand embraces those concepts at the designing level [3], merging compliance properties with a high-degree of underactuation. It has 19 degrees of freedom (DoF) actuated by a single motor which drives a tendon passing through the whole hand on an array of pulleys. As a result of the tendon routing, the hand behaves as a differential mechanism adjusting itself to different geometries and sizes.

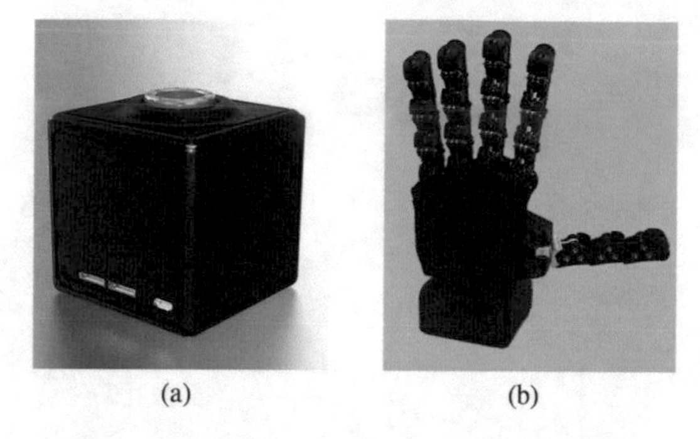

Fig. 3 (a) Qbmove Maker Pro VSA motors, (b) The Pisa/IIT SoftHand

3 Perception

A crucial point in picking operation is to define how the robot perceives the environment. Usually, information gathered from 2D or 3D sensors are collected and merged together, in order to provide the robot with data of surrounding objects and/or structures [14]. Although several strong improvements are achieved in past years, object recognition and object pose estimations are still two open problems.

In this work, information about the position, orientation, and other geometrical properties of the item, i.e., principal component, are collected. We used a RGB-D sensor assembled on the wrist of the robot.

Our algorithm extracts from a point cloud a group of clusters; then all of these are matched with the database of point clouds representing the items, finally, the best fit is selected as the target object. This approach is well suited to the challenge framework, where a fixed collection of objects was provided.

3.1 Point Cloud Database

The achievement of this technique requires to store a number of point clouds for each object. These data should be a thorough representation of the real object, and, in the meantime, avoid oversampling. Moreover, a fundamental role is played by the acquisition stage; indeed, to acquire coherent and meaningful data is essential in order to obtain good pose estimations.

We decided to discretize a full point cloud of an object in 36 faces; this process results in storing a point cloud every $\frac{\pi}{18}[rad]$. To acquire object poses, we used a rotating table powered by a servo-motor.

Point clouds collected are post-processed before being stored in the database; in particular, a segmentation algorithm isolates the object from the plane, and a roto-translation matrix is applied for expressing the object's point cloud in the robot reference frame, Fig. 4a.

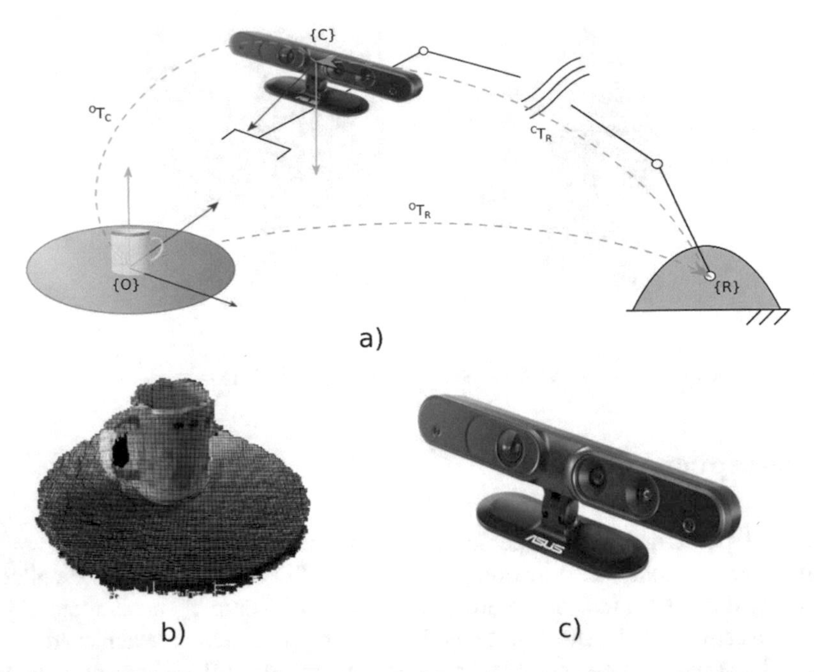

Fig. 4 (**a**) Robot-Camera transformation. $^{o}T_{c}$ and $^{c}T_{r}$ are the transformations of the object detected in the camera reference frame and the camera in the robot frame, respectively. Combining these transformations, the object pose is expressed w.r.t. the robot frame using $^{o}T_{r}$. (**b**) A sample pointcloud by the camera with the plane detected and coloured in red. The plane is removed before feeding the pointcloud into the ICP algorithm. (**c**) XtionPro 3D camera

3.2 Perception Algorithm

The core of our object recognition algorithm is a well-known methodology called *Iterative Closest Point (ICP)* [7]; it is a robust approach for matching two point clouds. However, it is extremely time-consuming, and applying it to each point cloud is very inefficient. Therefore, an initial filter stage selects just a subset of entries in the database which are likely to be the right match.

Taking advantage of objects' properties, we developed a filter for reducing the searching space.

The geometrical properties of an object roughly describe the shape of the object, and they can be very useful for pruning several possibilities in the database (Fig. 5).

Geometrically, every object can be described by its major components, namely height and width. In the database creation phase, our algorithm extracts this information for each item and then stores them in the database. The filter applies the same procedure to each input point cloud and compares the retrieved values with the ones in the database.

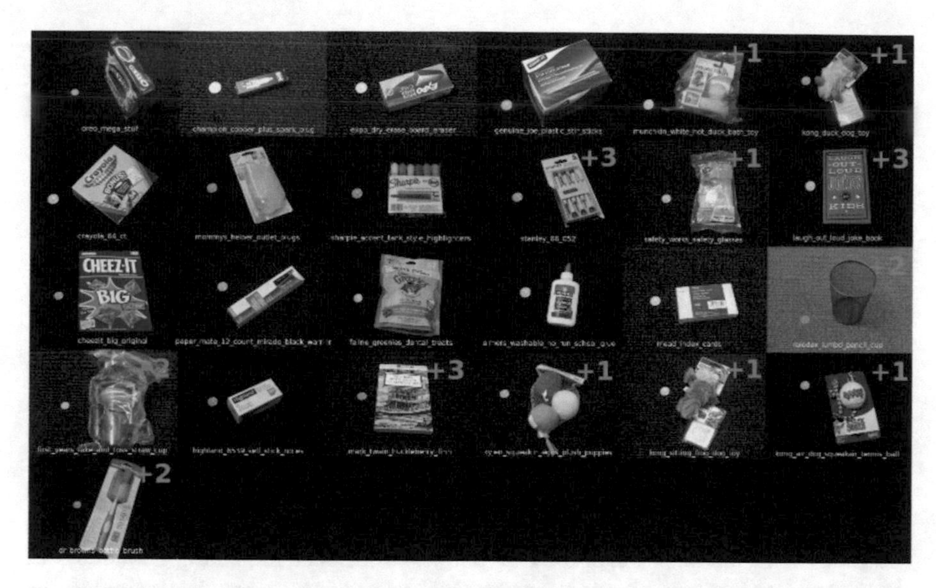

Fig. 5 Collection of 25 objects proposed for the APC

This approach fails when the perception system identifies more objects as a single item. However, since the rules set states that any object cannot prop on another during the challenge, the application of this filter showed reliable results in our experiments.

Thanks to this filter, we evaluate just a subset of the entire database improving the overall execution time of the recognition algorithm.

Before passing through the filtering process, an acquired new scene is segmented by an algorithm to remove all the outliers and isolates a single object. After these actions, the resultant query point cloud is passed to the *ICP*; at the end of the process, the point cloud in the database which best fits the requirements is selected as a target.

4 Motion Planning and Control

In the past decades, many approaches were proposed for planning the motion of a robot in grasping operation. Two macro groups of algorithms for this purpose are geometrical-based [15] and neural network-based [13]. The first exploits the knowledge of the environment for seeking the best configuration of contacts point on the surface of the object; then a planner generates a collision-free trajectory to the selected grasp. On the contrary, neural network-based algorithms take a point cloud as input and return grasp candidates at high likelihood based on a previous training set.

Both are well known in literature and lead to robust solutions with high performance in execution time and reliability. However, they look at the environment just as an obstacle to avoid, considering only the free space as searching area. Hence, this category of techniques does not allow to exploit *Environmental Constraints* (EC) [11] in the scene for improving the grasping success rate.

4.1 Motion Planner

We propose a soft approach for grasping, which tries to exploit ECs in order to facilitate and make robust and reliable grasping contact with a target object. In [10], a grasping planner has been proposed for exploiting such constraints. The authors divided the space into regions, depending on the present EC; then the algorithm builds a roadmap of actions interconnecting all regions. The key idea of this method is partially inspired by human that exploit the environment for achieving better grasps even in complex conditions. A paramount aspect for taking advantage of present constraints pass through a smart designing of the robot, in particular, the intrinsic softness of the selected hardware plays an essential role, and all approach would be unfeasible otherwise.

Figure 6 shows a schematic of our control system architecture. The entire system is divided in four blocks: *Grasp Strategies*, *Requested Movements*, *Trajectory Generator*, and *Control Loop*.

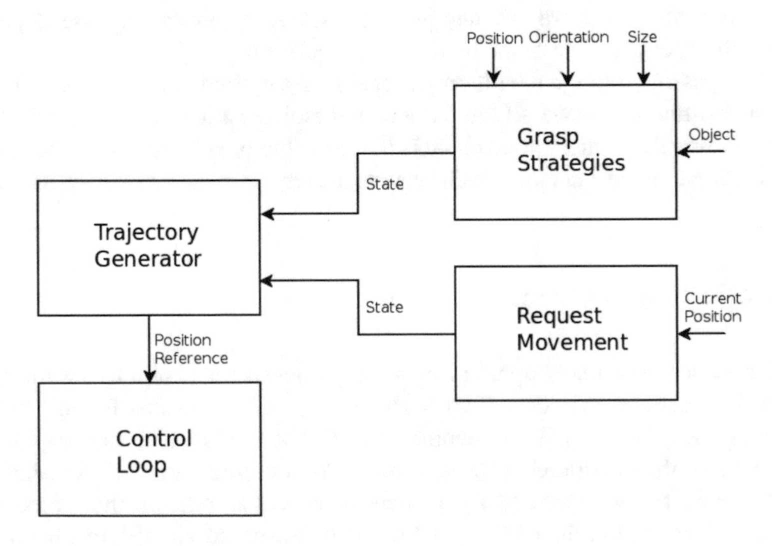

Fig. 6 Motion Planning flowchart

In most of the cases, for coping with EC exploitation, a single grasping operation is composed by a sequence of actions. In our method, *Grasp Strategies* and *Requested Movements* blocks provide inputs for generating the proper sequence of actions. The first selects the grasping strategy that most likely will allow a stable grasp, and the second generates motions in the e-e frame for approaching a target object.

Every action is described by a configuration state. Each state embeds a collection of information, the goal position of the joint values provided from the perception system, the amount of stiffness required for performing the operation which is defined by experimental observation and the hand state (open or closed). It is important to mark that the stiffness value defines how preciseness is required of the picking operation and it is extremely affected by the grasp chosen. After that a grasping generation is initialized; *Grasp Strategies* and *Requested Movements* generate a sequence of configuration states, which are gathered by the *Trajectory Generator* block. Then, these states are interpolated generating a trajectory from the path. The last stage of the algorithm, embraced by *Control Loop* block, executes the trajectory, in particular, PID controllers are designed for the three Cartesian Joints, and the wrist is made of VSA.

4.2 Grasping Strategies

The compliance property of the Pisa/IIT SoftHand is provided by the synergetic actuation and the tendons which permit fingers to adjust themselves upon the surface to be grasped. This feature broadly improves the grasping potential, keeping intrinsically simple the whole application [4]. Secondly, information gathered by the camera is hardly thorough; in fact, it is very common to collect uncertainty data, due to noise-like refractions. The adaptability property of the Pisa/IIT SoftHand, in many cases, overcomes those issues allowing to bump against the target object; indeed the hand can be warped without any damage to itself or external elements.

However, some downsides come with an increased complexity in the approaching stage.

Classic grasping approaches analyze features of the object, like shape and size, to extrapolate the minimum group of points which yields the best grasp in term of stability or effort, i.e., force-form closure [12]. Those techniques demand a good knowledge of the object to grasp; thus, they can be very challenging to apply in real scenarios, where gathering full information and time-consuming processes are despicable. Therefore, we devised a set of grasp strategies composed of relative motions, and each of them is designed for taking advantage of the softness properties of the robot during the exploitation of the present ECs in the scene. As a result of its designing, the Pisa/IIT SoftHand acts very similar to a human hand in the phase of approaching and grasping an object. Because of that, it would be very easy to collect data by human demonstration experiments that can lead to efficient results when copied by a robot.

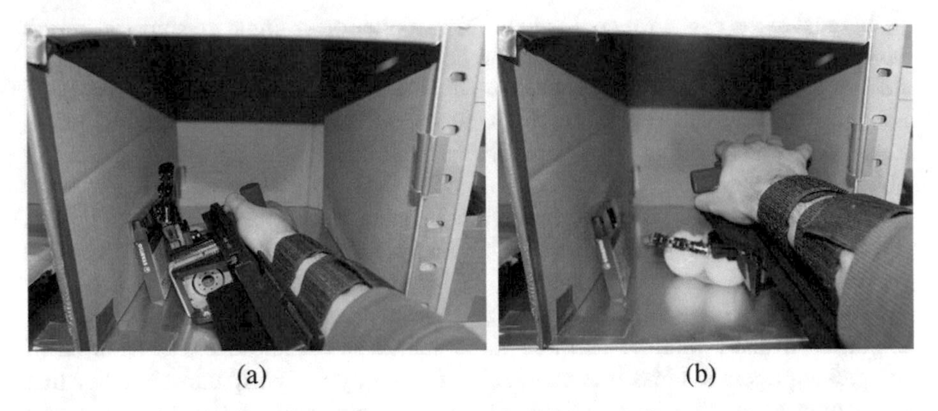

Fig. 7 (**a**) Pick up an object from Left Side, (**b**) Pick up an object from Center

Our approach to this problem is using a wearable hand; thanks to it a human operator can substitute its hand with the robotic hand, Fig. 7. The device is buckled up to the user's forearm by straps, then, via a lever, the operator can modulate the gripper closure.

We have divided a bin in three zones: left side, middle, and right side. Because of present constraints, like walls, to operate in a specific sector can decrease the feasible space for the hand, i.e., pick up an item along the left side does not allow all possible orientations.

The experimental setup involved a human operator, who, after a training phase, performed a picking task of all objects over all sectors.

These experiments lead to the definition of $\approx$30 different grasping strategies; however, a further categorization has decreased this number to 16, explained in more detail in the next section.

All data collected were stored in a database, where at each point cloud is associated a grasping strategy. Therefore, our algorithm queries the database at runtime for retrieving the best method for grasping based on human demonstration. Observing the results of the hand-in-hand experiments, a correlation between some strategies was enlightened by a high frequency of some features, like gripper pose displacement and wrist configuration.

Thanks to this correlation, we have pruned all strategy to sixteen macro-classes. The remaining strategies are independent of one another and represent a more general set of approaches. However, some strategies involve more than an action, thus some connection between classes was created to take into account those cases.

In Fig. 8 are depicted the final sixteen classes. Every big circle represents a macro-class, and each little circle connected to it is a sequence of action.

As previously mentioned, the *trajectory generator* block generates a trajectory using the sequence of states (actions) in the queue. For interpolating two adjacent state vectors, the algorithm adopts a sigmoid-based approach following Eq. (1).

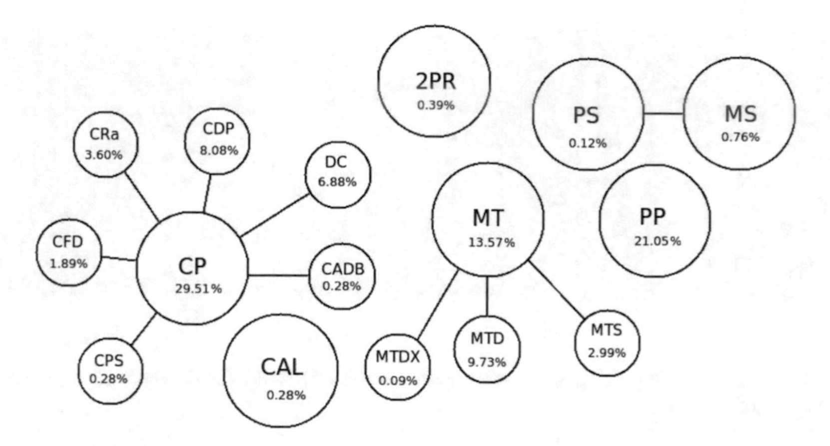

Fig. 8 Grasp Strategies implemented

The resulting interpolation respects constraints on the goal time and position, maintaining a sigmoid-like shape of the joint values w.r.t. time.

$$y_i(k) = y_i(k_0) + \frac{e_i(k_0)}{1 + \exp(-0.3\sigma(k))}, \qquad \forall i \in \{1 \rightarrow 6\} \tag{1a}$$

$$\sigma(k) = -15 + \frac{30k}{max(e)}\hat{v} \tag{1b}$$

Where $e \in \Re^6$ is the position error vector of each motor, $y \in \Re^6$ is the current reference for each motor, $[k_0, k]$ is the window of time, Δt is the time step, and $\hat{v}$ is the desired velocity. The planner also provides the robot with a coefficient, which represents the required closure rate of the Pisa/IIT SoftHand; its values are within the range [0, 1], where 0 means entirely open and 1 fully closed.

5 Conclusions and Lesson Learned

In this work, a novel approach, based on soft robotics, is proposed to automate picking of objects with different shapes, in cluttered and constrained environments (Fig. 9).

Two key devices of our robot are the Pisa/IIT SoftHand, a robust and adaptive anthropomorphic hand, and the Qbmoves, VSA modular actuators used to compose the robot wrist.

The planning and control architecture lean on a set of grasp strategies resulting in a collection of hand-in-hand experiments performed by a human pilot grasping

Fig. 9 (**a**) Example of picking up an object with Pisa/IIT SoftHand. (**b**) Example of picking up an object with the piston.

objects with the Pisa/IIT SoftHand thanks to a handle. The grasp strategies are categorized into 16 classes; then those strategies are divided into three macro-classes depending on the object's location inside the bin.

A perception system, based on a 3D camera, is used to perform an object's pose (position and orientation) recognition. On the basis of the outputs coming from the perception system, namely the type of object and its position, a suitable grasp strategy is triggered.

The resulting experiments reveal good performance, and high reliability on grasping single objects, also under wide uncertainties on the object's position and orientation. This mainly thanks to the adaptability property inherent of the robot structure.

When the target object is in a cluttered environment, the success rate substantially decreases because it is necessary to push away other objects in order to clear the scene. This operation is very challenging and hardly automatable. Moreover, we observed that the perception system is strongly affected by the brightness, and it lacks accuracy and reliability when surrounding conditions change. A video of the robot during the challenge is available on [9].

References

1. Amazon Picking Challenge rule (2015). http://pwurman.org/amazonpickingchallenge/2015/detailsrules.html
2. Angelini, F., Santina, C.D., Garabini, M., Bianchi, M., Gasparri, G.M., Grioli, G., Catalano, M.G., Bicchi, A.: Decentralized trajectory tracking control for soft robots interacting with the environment. IEEE Trans. Robot. **34**, 924–935 (2018)
3. Bicchi, A., Gabiccini, M., Santello, M.: Modelling natural and artificial hands with synergies. Philos. Trans. R. Soc. B Biol. Sci. **366**, 3153–3161 (2011)
4. Bonilla, M., Farnioli, E., Piazza, C., Catalano, M., Grioli, G., Garabini, M., Gabiccini, M., Bicchi, A.: Grasping with soft hands. In: IEEE-RAS the International Conference on Humanoid Robots (2014)

5. Catalano, M.G., Grioli, G., Garabini, M., Bonomo, F., Mancini, M., Tsagarakis, N.G., Bicchi, A.: VSA – CubeBot. A modular variable stiffness platform for multi degrees of freedom systems. In: IEEE International Conference on Robotics and Automation, Shangai, China (2011)
6. Catalano, M.G., Grioli, G., Serio, A., Farnioli, E., Piazza, C., Bicchi, A.: Adaptive synergies for a humanoid robot hand. In: IEEE-RAS International Conference on Humanoid Robots (2012)
7. Chen, Y., Medioni, G.: Object modelling by registration of multiple range images. Image Vision Comput. **10**(3), 145–155 (1991)
8. Della Santina, C., Piazza, C., Gasparri, G.M., Bonilla, M., Catalano, M.G., Grioli, G., Garabini, M., Bicchi, A.: The quest for natural machine motion: an open platform to fast-prototyping articulated soft robots. IEEE Robot. Autom. Mag. **24**(1), 48–56 (2017)
9. E. Piaggio Research Centre Team Highlights at APC (2015). https://www.dropbox.com/s/s2kdf8914cczvol/AmazonPickingChallenge2015-Highlights.avi?dl=0
10. Eppner, C., Brock, O.: Planning grasp strategies that exploit environmental constraints. In: IEEE the International Conference on Robotics and Automation (2015)
11. Eppner, C., Deimel, R., Álvarez-Ruiz, J. Maertens, M., Brock, O.: Exploitation of environmental constraints in human and robotic grasping. Int. J. Robot. Res. **34**, 1021–1038 (2015)
12. Ferrari, C., Canny, J.: Planning optimal graphs. In: IEEE/RSJ the International Conference on Robotics and Automation (1992)
13. Gualtieri, M., ten Pas, A., Saenko, K., Platt, R.: High precision grasp pose detection in dense clutter. In: IEEE International Conference on Intelligent Robotics (2016)
14. Izadi, S., Kim, D., Hilliges, O., Molyneaux, D., Newcombe, R., Kohli, P., Shotton, J., Hodges, S., Freeman, D., Davison, A., Fitzgibbon, A.: KinectFusion: real-time 3D reconstruction and interaction using a moving depth camera. In: ACM Symposium on User Interface Software and Technology (2011)
15. Miller, A., Allen, P.K.: Graspit!: A versatile simulator for robotic grasping. IEEE Robot. Autom. Mag. **11**, 110–122 (2004)
16. Rus, D., Tolley, M.T.: Design, fabrication and control of soft robots. Nature **521**, 467–475 (2015)
17. Santello, M., Flanders, M., Soechting, J.F.: Postural hand synergies for tool use. J. Neurosci. **18**, 10105–10115 (1998)
18. Vanderborght, B., Albu-Schaeffer, A., Bicchi, A., Burdet, E., Caldwell, D.G., Carloni, R., Catalano, M.G., Eiberger, O., Friedl, W., Ganesh, G., Garabini, M., Grebenstein, M., Grioli, G., Haddadin, S., Hoppner, H., Jafari, A., Laffranchi, M., Lefeber, D., Petit, F., Stramigioli, S., Tsagarakis, N.G., Van Damme, M., Van Ham, R., Visser, L.C., Wolf, S.: Variable impedance actuators: a review. Robot. Auton. Syst. **61**, 1601–1614 (2013)

Top-Down Approach to Object Identification in a Cluttered Environment Using Single RGB-D Image and Combinational Classifier for Team Nanyang's Robotic Picker at the Amazon Picking Challenge 2015

Albert Causo, Sayyed Omar Kamal, Stephen Vidas, and I-Ming Chen

1 Introduction

In this paper, we are interested in the problem of fully automated picking of items from an arbitrarily stocked shelving unit. The robot must pick the intended item from the shelf based on a single RGB-D image of the shelf bin. In order to determine proper grasp strategy for the robotic picker, proper identification of the target item in a cluttered environment is necessary.

Most robotic picking problems consider an open environment with objects placed on a planar surface without clutter or occlusion. Our robotic picker deals with objects placed in a shelf, where the walls of the shelf limit the possible viewing angles of the objects and clutter from multiple objects is present due to the confined space. While some methods are able to solve the problem of object identification in controlled environments, reliable object identification in uncontrolled or poorly controlled environments remains elusive [1, 2], particularly for arbitrarily posed objects in cluttered and/or occluded environments. The main concerns of such systems include practicality, speed, and robustness of identification.

A. Causo (✉) · I.-M. Chen
School of Mechanical and Aerospace Engineering, Nanyang Technological University, Singapore, Singapore
e-mail: acauso@ntu.edu.sg; michen@ntu.edu.sg

S. O. Kamal
Transforma Robotics, Singapore, Singapore
e-mail: sokamal@transformarobotics.com

S. Vidas
School of Electrical and Electronics Engineering, Nanyang Technological University, Singapore, Singapore
e-mail: svidas@ntu.edu.sg

A. Causo et al. (eds.), *Advances on Robotic Item Picking*,
https://doi.org/10.1007/978-3-030-35679-8_4

A classical approach to object recognition is to consider the color, texture, or other invariant features of the object as a differentiator [3]. Commonly used in histograms, color has proven to be an important cue in the identification of objects. Another popular approach is the use of edge based features, where the geometry of an object is used either in its entirety or as collections of fragments [2]. To assist in identification and/or segmentation, features are also tagged with shape descriptors [4], parametric surface descriptions [5], and shape and contour primitives [6]. Feature combination of different modalities such as geometric and texture based features is also a common strategy [7].

Patch based features are also widely used for recognition, where characteristic boundaries that sufficiently describe the features of objects, either in patches of rectangular shapes (local features) or irregular patches (region based features) are used. Popular algorithms involving patch based features include SIFT [3], SURF [8], FAST [9], and BRISK [10]. A newer algorithm which boasts substantially good performance on par with SIFT and SURF is ORB (Oriented FAST, Rotated BRIEF) [11].

Current research makes use of probabilistic modeling to learn certain features of the observed data to identify object over a broad range of conditions. It allows the recognition to be robust to uncertainty and noise and adaptive and scalable to large datasets. Two main classifications exist for the learning process: bottom-up (discriminative) and top-down (generative) approaches. Bottom-up approaches take low- and/or mid-level features to build hypotheses and extend them by construction rules before evaluating them through cost functions. Top-down approaches obtain class-specific model features or object configurations, usually from training images, so as to obtain hypotheses from models matched to images [12]. The approaches differ in the flow of information, where bottom-up moves from the observed image to the state variables while top-down moves in the opposite direction.

This paper presents an approach to object identification that combines feature-based methods to create a strong classifier by balancing the strengths and weaknesses of the individual classifiers. Using a single markerless RGB-D image obtained from a Microsoft Kinect sensor, all objects within the image are segmented into separate blobs (segments) for feature extraction. The features (properties) are then compared to a set of trained object data so as to predict the probability of a correct labeling of the target item to the appropriate blob. We use a combination of HSV histograms, blob contours, and ORB features classifiers to create a top-down probabilistic model of each item and their various possible poses within the shelf bin.

2 The Picking Problem

The robot will use the proposed vision approach on the picking problem based on the Amazon Picking Challenge in 2015, where an Amazon shelf with 12 bins is loaded with up to 25 different items. Figure 1 shows the different items while Fig. 3 shows

Fig. 1 The 25 items used for the picking challenge (image taken from the Amazon Picking Challenge website [13])

the shelf's close-up. The items differ in weight (30–500 g), shape (soft toys with organic shape to boxes), color (transparent, white, colorful, or black), and packaging (plastic wrapped, boxed, or even no packaging). Some of the items are 64-color Crayola, Cheez-It, Dr. Brown Bottle Brush, a spark plug, glue, board eraser, self-stick notes, books, index cards, toys such as squeaking eggs plush puppies, rubber duckies, and tennis ball. The complete list of the items is also tabulated in Fig. 13.

The robot has to pick an item from each bin containing between 1 and 5 items, including the target item, arranged randomly. The shelf has 12 bins and the front face measures 0.90 m × 0.865 m. Each bin measures either 0.25 m × 0.42 m × 0.25 m or 0.30 m × 0.42 m × 0.25 m (width × depth × height). The bottom bins are 0.78 m from the floor (as shown in Fig. 2). There are 25 items to be picked in total. The items are products from amazon.com that include a pack of Oreo, spark plug, board eraser, a box of plastic stir sticks, duck bath toy, 64-piece Crayola, outlet plugs, a pack of Sharpies, screw driver set, a box of Cheez-It, a box of pens, cat food, whiteboard eraser, Elmer's glue, a pack of index cards, pencil cup holder, plastic drinking cups, sticky notes, a couple of plushy toys, and a couple of books. The items come in various shapes and packaging. There are boxed or box-like items, like the crayons, box of stir sticks, and the books, while others come in loose plastic packaging, like the plushy toys and plastic drinking cups. Still others do not have any packaging at all, like the rubber duckie and pencil cup holder (Fig. 1).

Two lists are provided at the beginning of each picking session: the first one is the list of the target items (i.e., items to pick) and the second is the list of all items in every bin, which includes the target item. The robot has to pick and place the items into the order bin autonomously within 20 min. Each item picked correctly, and without damage to its packaging, is given a corresponding score.

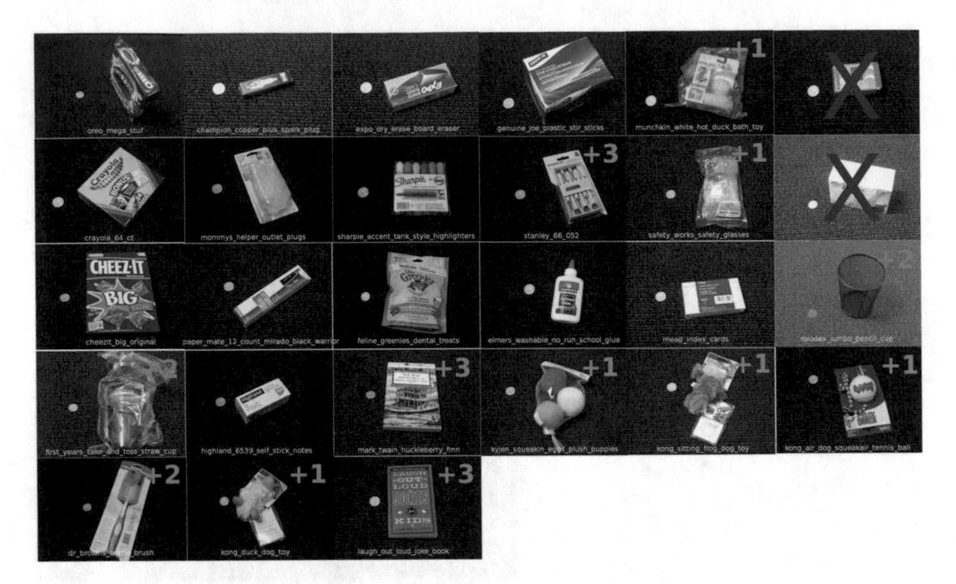

Fig. 2 The robotic picking system deployed during the Amazon Picking Challenge 2015 [13]. The shelf has 12 bins in total

Fig. 3 A close-up of the Mantis gripper as it picks an item from the shelf. For this item, the gripper deployed both its suction and its parallel jaw gripper. The Kinect sensor is mounted at the end-effector, just above the gripper

3 The Robotic Picking System

The robotic picking system is composed of a robot arm (Universal Robot's UR5 model) with six degrees of freedom, an RGB-D camera, and a specially designed gripper. The robot is mounted on a pedestal in order to reach the top and bottom bins of the shelf (see Fig. 2). A Kinect camera augmented with externally mounted lights acts as the vision system. We used a custom-designed gripper nicknamed Mantis which has two suctions and a parallel jaw gripper that allow for different grasp strategies. The gripper is shown in Fig. 3. A program manager coordinates the vision, grasping, and task planning modules of the robot to accomplish its picking task. The whole system is calibrated (including the shelf, camera, and grippers) with respect to the robot's axis.

4 Top-Down Identification Method

Our proposed top-down identification method begins with the captured RGB-D image data from the Kinect sensor. The data is obtained by moving the robot "head" (the camera and the gripper) right at the front of the target bin (the bin where an item will be picked) and image is captured. The area within the image which contains the target bin is labeled as the region of interest (ROI). Canny edge detection is run on the ROI to segment and identify the possible target object. With the edge as guide, blobs are identified within the ROI. Further morphological processing (dilation and erosion) are run on the blobs to obtain N blobs, where N is the number of items inside the bin, as given in the list provided by the system at the beginning of the picking run (see Sect. 2). Refer to Fig. 4 for a sample ROI and blob identification.

Each segmented blob in the RGB-D image represents an unidentified object, which is then described through a set of extracted global features as discussed in Sect. 4.1. Each blob is then checked against the known target item and any other items within the shelf bin, which is known from the picking order list. The blob that returns the highest probability match with the target item is then identified as the "correct" item and its pose estimated by extracting the point cloud of the blob and comparing it against a library model point cloud of the target item. Such feature extraction is performed both for the training data captured for the purpose of generating the probabilistic model (described in Sect. 4.2).

4.1 Feature Extraction

Extracted global features for each segmented object include a hue-saturation histogram, geometric contour, visual BRISK features, and the estimated dimensions of the object in millimeters. Each global feature can be stored as a set of arrays, e.g., for the histogram there is one array for each of the hue and saturation channels as shown in Fig. 5.

For this work, the number of different global features G was limited to 4. The matching distance d_g (where $g \in G$) between corresponding global features extracted from two separate image segments is calculated using standard methods

Fig. 4 Segmented objects from a typical image of the shelf bin. The blue rectangle denotes the region of interest (ROI) and the red and green contours display the segmented blobs, one of which could be the possible target object that initially have unknown identity

which vary depending on the nature of the feature, but achieve the same result of a lower distance indicating more similarity between the segments.

4.2 *Probabilistic Modeling*

A probabilistic model is generated offline which can be used to enhance the accuracy of the system as it is applied online to unseen data. First, a library of global features is generated by extracting features as described above from images taken that cover a range of possible viewpoints of each object such as shown in Fig. 6.

Next, independently captured and annotated test data is collected, where the correct class identity and best matching viewpoint in the library is recorded. For each of the global features, the matching distance d_g between each test segment and the best matching segment from each object class is stored. For a single global feature, these best (lowest) distances will be denoted as D_n where $n \epsilon N$, with n representing the index of the object class out of N possible indices.

Then, for each test segment, a set of $N - 1$ ratios are calculated. For each ratio, the numerator is D_m (where m is the index of the correct object class), while the denominator is D_n for all possible values of n except for $n = m$.

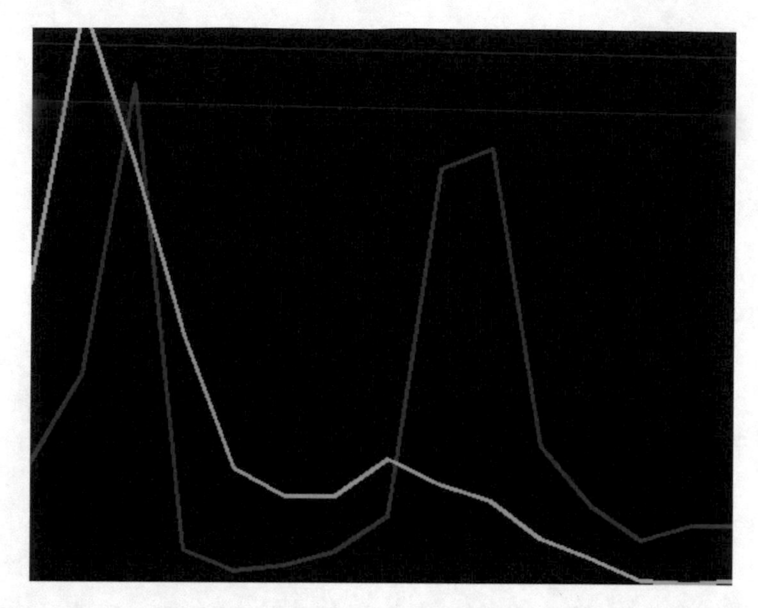

Fig. 5 Color-histogram global feature represented as two curves. The blue curve represents the hue, ranging from 0 to 180 along the x-axis. The green curve represents the saturation, ranging from 0 to 255 along the x-axis. The y-axis is scaled to normalize the area under each curve for display purposes. Every different segment has a different histogram; however, segments that represent the same object under similar viewpoints are likely to have similar histograms and, therefore, achieve lower (better) matching scores

Fig. 6 Two variations of a single object (a bottle of glue). In this case, the two images that are included in the image library differ both in the pose of the object (laying flat versus standing), and the particular version of the object (opaque versus translucent)

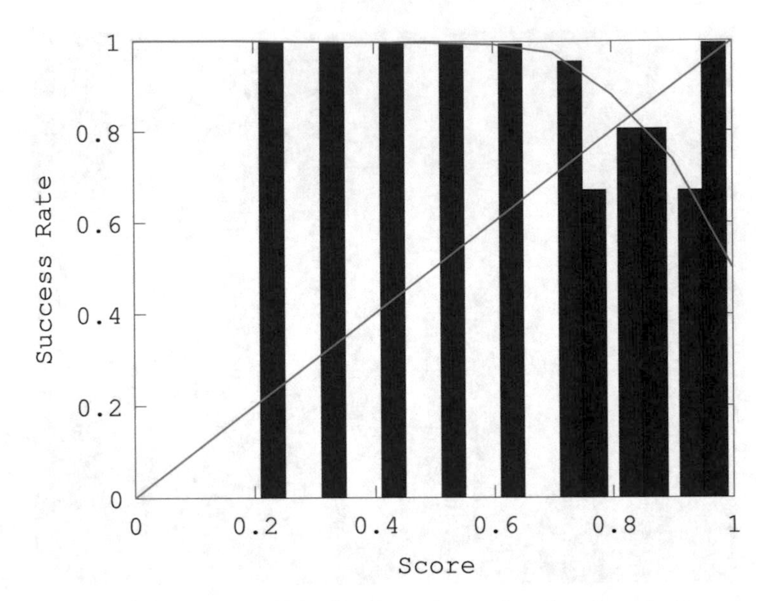

Fig. 7 Histogram of success rate of identification estimates based on the color-histogram global feature. Each blue column shows the proportion of tests that were successful (y-axis) where the matching ratio was approximately the value on the x-axis. The red column shows the results when these ratios are converted to probabilities by mapping the fitted blue curve to the ideal red curve

Therefore, the various numerators will contain all of the matching distances for incorrect classes. From a histogram representing the success rates of various ranges of these matching ratios $\frac{D_m}{D_n}$ (shown in Fig. 7), a smooth mapping curve is then estimated that converts a matching ratio into a probability. Generally ratios closer to 0 will have probabilities closer to 1, and the probability will converge to around 0.5 as the ratio gets closer to 1, indicating that the best and second-best labels have a similar level of similarity with the test object, and therefore a similar likelihood.

5 Results and Discussion

Figure 8 shows our proposed identification algorithm running. Figure 8a shows the raw image captured by Kinect. The algorithm identifies the book (green contour in (b)) among the three objects in the bin while the segmented masks are shown in (c) and the estimated alignment of the obtained point cloud to the library model in (d). The red point cloud shows the library model cloud, the blue point clouds show the multiple assumptions of the target cloud, and the orange point cloud shows the best matching point cloud orientation.

The training process captures image data of all objects at various rotations and orientations to generate full features at all possible orientations. Figure 9 shows a

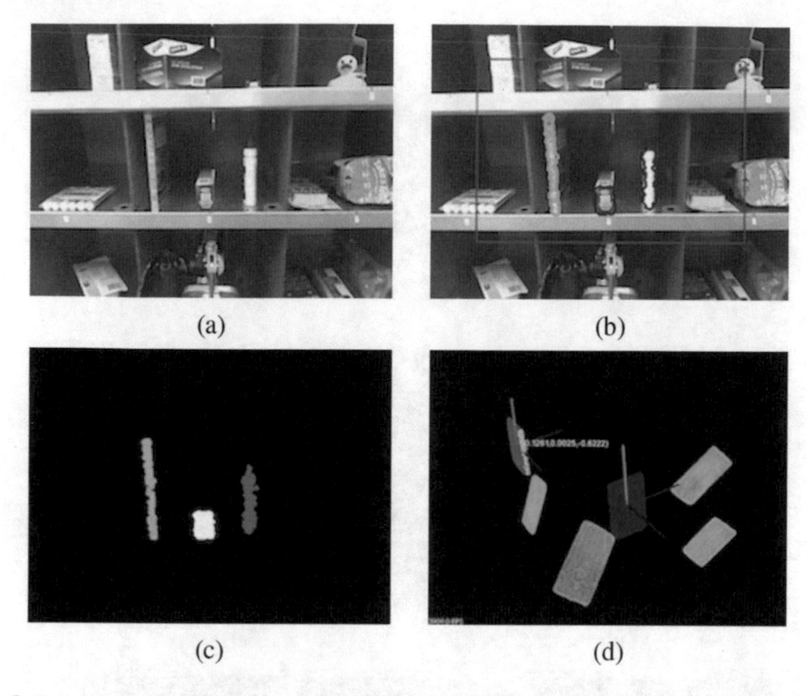

Fig. 8 Raw image (**a**); ROI and segmented blobs identified (**b**); segmented point cloud data based on the result of identification (**c**); and pose estimate of the point cloud data (**d**)

sample of the training run using *Cheez-It big original* and *first years take and toss straw cup* as examples.

5.1 Image Identification

The offline probabilistic modeling is condensed into a 25×25 matrix we call the identification prior probability matrix (IPPM) (Fig. 13). The IPPM gives the system an initial estimate of the likelihood of confusion between the multiple objects when they are placed in the same shelf bin. It is the quantification of the combinational classifiers obtained at the end of the training and validation phase.

A value of 1 means that the item is estimated to have no chance of being mislabeled as another item. Probabilities less than 1 indicate the likelihood of the object being mislabeled as each other. For example, from Fig. 13, the *Laugh Out Loud joke book* (item number 15) has zero (0) chance of being mislabeled as *Cheez-It big original* (item number 2) but has a $(1 - 0.47)$ chance of being mislabeled as *champion copper plus spark plug* (item number 1).

The identification algorithm performs well under multi-item and single item identification as shown in Figs. 8 and 10, respectively. Initially, the algorithm

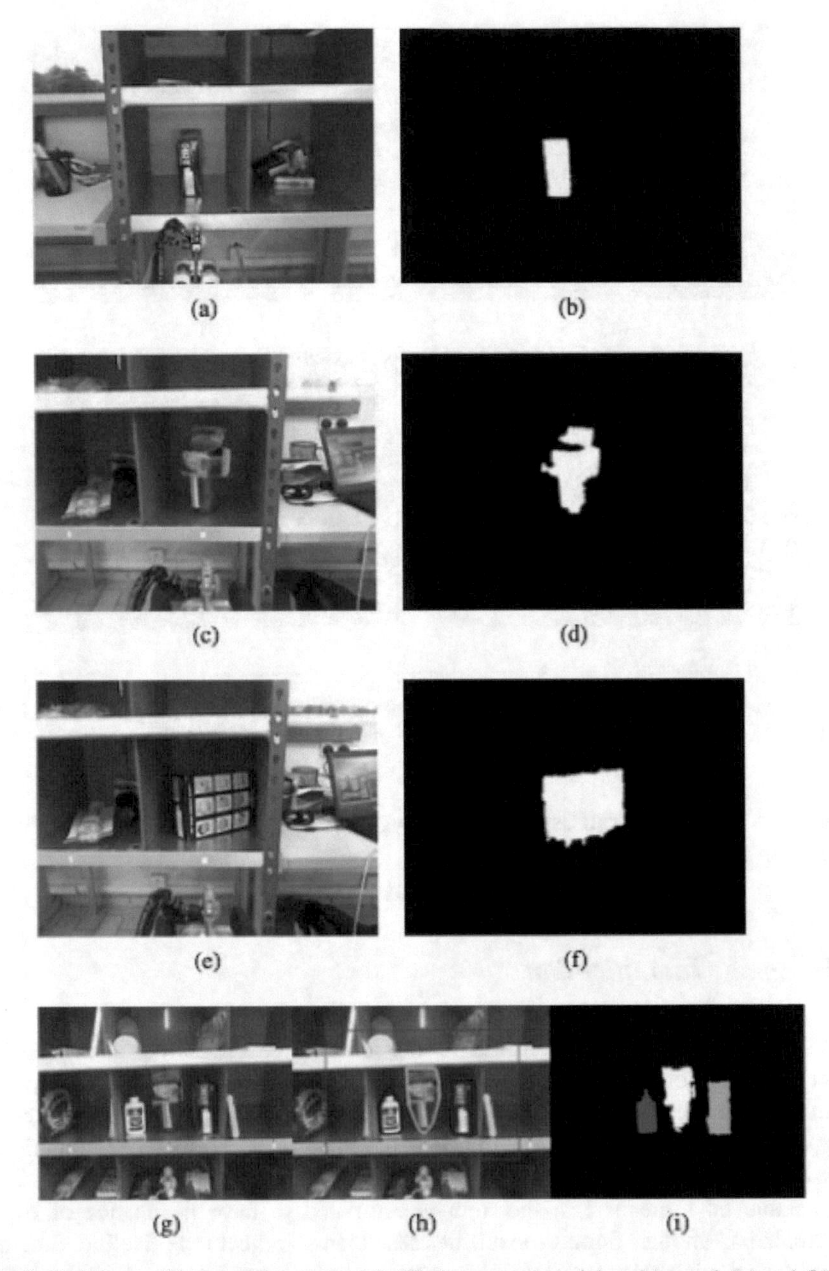

Fig. 9 Training, validation and test sets. (**a**)–(**d**) show sample training data used. (**a**) and (**b**) are the RGB and segmented object mask for the *Cheez-It big original* (depth image is also used but is not shown here as it would simply show white blobs) and (**c**) and (**d**) for the *first years take and toss straw cup*. (**e**) and (**f**) show a validation image for the *Cheez-It big original*, where the same object is compared to itself at the different orientations to establish how similar/dissimilar the images are. (**g**)–(**i**) show sample test data where the target object *first years take and toss straw cup* is placed in a bin together with other objects, in this instance, the *Cheez-It big original* and *Elmer's washable no run school glue*, in random arrangement and the system is tasked with finding the target among the items

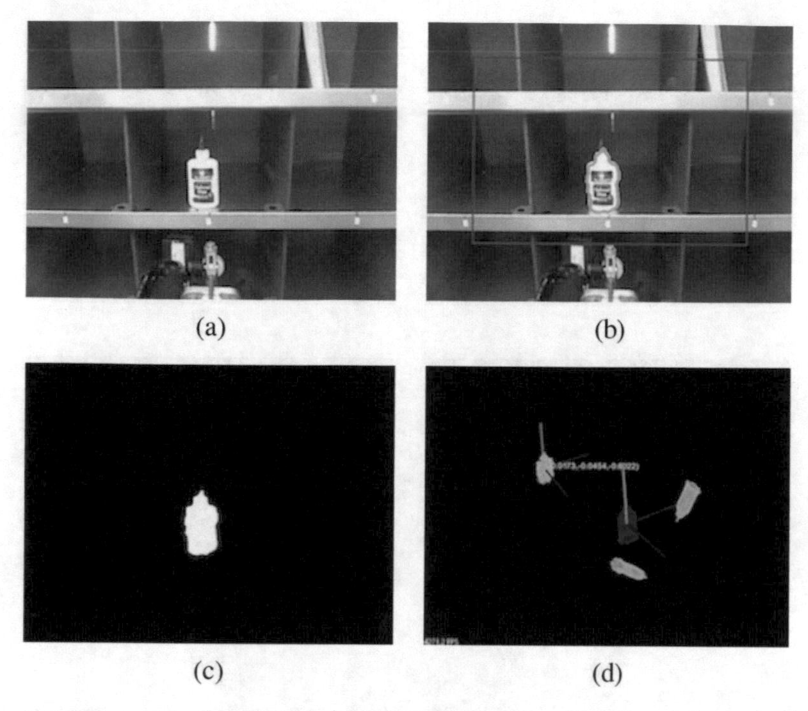

Fig. 10 Identification algorithm performed on single item in the shelf bin. (**a**) Raw camera image. (**b**) The item segmented from the input image with its bounding box shown. (**c**) The binary image of the segmented item. (**d**) Pose estimate of the segmented item

segmentation captures a rough image patch as seen in the above mentioned figures but with refinement, the image patches can be made to capture a smoother image patch as shown in Fig. 11.

Some of the challenges of the algorithm include the wrong segmentation of objects from RGB-D images. Due to the nature of depth sensing on surfaces not directly facing the sensor, the segmentation result could end up wrongly as depth could not be perceived, resulting in single objects being segmented as multiple objects, as seen in Fig. 11. Another challenge we have encountered is occlusion. The current algorithm is not very robust to occlusion including when adjacent objects are just too close (touching) to each other. Occlusion results in multiple objects segmenting into a single object, as seen in the examples in Fig. 11. Also, some items are just too difficult to detect due to packaging or component material. For example, the *Rolodex jumbo pencil cup*, a pencil holder made with metal mesh, shown in Fig. 12. The problem is that the cup itself cannot be detected by the Kinect sensor as the wire mesh is too thin a structure that no depth data is perceived, thus the segmentation ignores it completely (Fig. 13).

Figure 14 shows additional result where the system is able to find the correct target and compute its 3D position. The bottom row also shows that our approach could differentiate and identify an item even if it is tilted instead of being in an upright position.

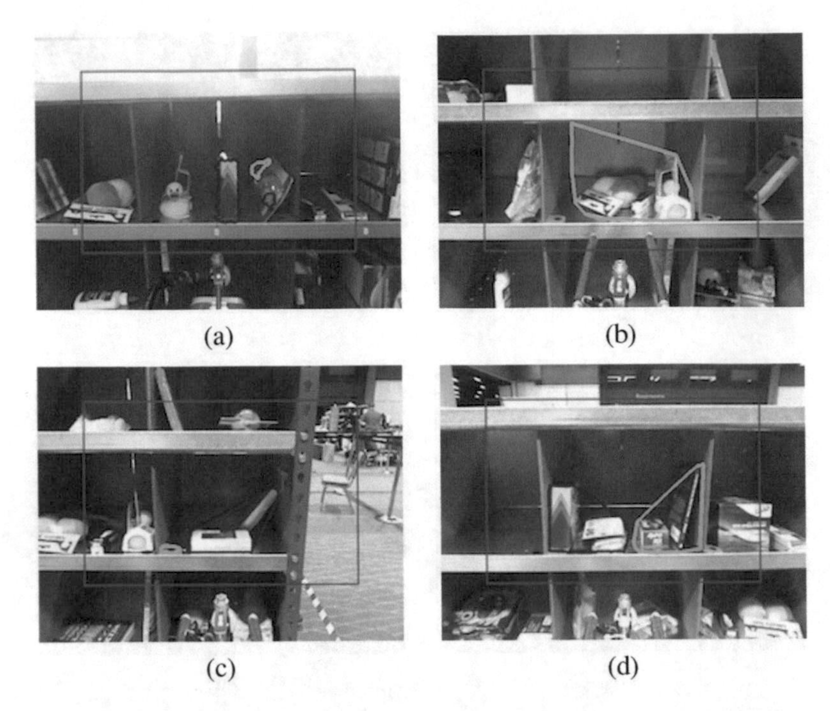

Fig. 11 Some of the problems faced by the system. Depth problem where incomplete depth sensing results in single-to-multi object segmentation (**a**). Occlusion problem where multi-to-single object segmentation occurs (**b**)–(**d**)

Fig. 12 The non-detection of the *rolodex jumbo pencil cup* results in bad detection. (**a**) The book is shown as the target item instead of the pencil cup on its left. (**b**) The segmented and binarized image shows the book instead of the cup

	1	2	3	4	5	6	7	8	9	10	11	12	13	14	15	16	17	18	19	20	21	22	23	24	25
1:champion_copper_plus_spark_plug	1	1	1	1	1	1	1	1	1	1	1	1	1	1	0.47	1	1	1	1	1	1	1	1	1	1
2:cheezit_big_original	1	1	1	1	1	1	1	1	1	1	1	1	1	1	1	1	1	1	1	1	1	1	1	1	1
3:crayola_64_ct	1	1	1	1	1	1	1	1	1	1	1	1	1	1	1	1	1	1	0.99	1	1	1	1	1	1
4:dr_browns_bottle_brush	1	1	1	1	1	1	1	1	1	1	1	1	1	1	1	1	1	1	1	1	1	1	1	1	1
5:elmers_washable_no_run_school_glue	1	1	1	1	1	1	1	1	1	1	1	1	1	1	1	1	0.96	1	1	1	1	0.99	1	1	1
6:expo_dry_erase_board_eraser	1	1	1	1	1	1	1	1	1	1	1	1	1	1	1	1	1	1	1	1	1	1	1	1	1
7:feline_greenies_dental_treats	1	1	1	1	1	1	1	1	1	1	1	1	1	1	0.99	1	1	1	1	1	1	1	1	1	1
8:first_years_take_and_toss_straw_cup	1	1	1	1	1	1	1	1	1	1	1	1	1	1	1	1	1	1	1	1	1	1	1	1	1
9:genuine_joe_plastic_stir_sticks	1	1	1	1	1	1	1	1	1	1	1	1	1	1	1	1	1	1	1	1	1	1	1	1	1
10:highland_6539_self_stick_notes	1	1	1	1	1	1	1	1	1	1	1	1	1	1	1	1	1	1	1	1	1	1	1	1	1
11:kong_air_dog_squeakair_tennis_ball	1	1	1	1	1	1	1	1	1	1	1	1	1	1	1	1	1	1	1	1	1	1	1	1	1
12:kong_duck_dog_toy	1	1	1	1	1	1	1	1	1	1	1	1	1	1	1	1	1	1	1	1	1	1	1	1	0.99
13:kong_sitting_frog_dog_toy	1	1	1	1	1	1	1	1	1	1	1	1	1	1	1	1	1	1	1	1	1	1	1	1	1
14:kyjen_squeakin_eggs_plush_puppies	1	1	1	1	1	1	1	1	1	1	1	1	1	1	1	1	1	1	1	1	1	1	1	1	1
15:laugh_out_loud_joke_book	0.47	1	1	1	1	1	0.99	1	1	1	1	1	1	1	1	0.94	1	1	1	1	1	1	1	1	1
16:mark_twain_huckleberry_finn	1	1	1	1	1	1	1	1	1	1	1	1	1	1	0.94	1	1	0.99	1	1	0.98	0.99	0.98	0.99	1
17:mead_index_cards	1	1	1	1	0.96	1	1	1	1	1	1	1	1	1	1	1	1	0.85	0.97	1	1	1	1	1	1
18:mommys_helper_outlet_plugs	1	1	1	1	1	1	1	1	1	1	1	1	1	1	1	0.99	0.85	1	1	1	0.99	0.98	0.97	1	1
19:munchkin_white_hot_duck_bath_toy	1	1	0.99	1	1	1	1	1	1	1	1	1	1	1	1	1	0.97	1	1	1	1	0.99	1	1	1
20:oreo_mega_stuf	1	1	1	1	1	1	1	1	1	1	1	1	1	1	1	1	1	1	1	1	1	1	1	1	1
21:paper_mate_12_count_mirado_black_warrior	1	1	1	1	1	1	1	1	1	1	1	1	1	1	1	0.98	1	0.99	1	1	1	1	1	1	1
22:rolodex_jumbo_pencil_cup	1	1	1	1	0.99	1	1	1	1	1	1	1	1	1	1	0.99	1	0.98	0.99	1	1	1	0.96	1	1
23:safety_works_safety_glasses	1	1	1	1	1	1	1	1	1	1	1	1	1	1	1	0.98	1	0.97	1	1	1	0.96	1	1	1
24:sharpie_accent_tank_style_highlighters	1	1	1	1	1	1	1	1	1	1	1	1	1	1	1	0.99	1	1	1	1	1	1	1	1	1
25:stanley_66_052	1	1	1	1	1	1	1	1	1	1	1	0.99	1	1	1	1	1	1	1	1	1	1	1	1	1

Fig. 13 The identification prior probability matrix (IPPM) for the 25 items listed in the Amazon Picking Challenge

Fig. 14 Identification algorithm identifying correct target items and converting the 2D positions into real-world coordinates for the robot picking task. (**a**)–(**d**) show the algorithm performing on several item arrangements. (**e**)–(**h**) show the system's ability to differentiate between standing pose and tilted pose

5.2 Competition Performance

The official score of the team was 11 (one item was correctly picked) as the system suffered from a software glitch at the beginning of its run, which also shorten the allotted time to finish the task from 20 min to roughly 7 min. However, the judge gave an unofficial time extension of 14 more minutes just for us to see what would have happened if we had the full 20 min. The unofficial score of the team was 117 (successful picking of 8 items out of 12). Failure to pick the four items was due to failure to properly grasp the product due to its packaging (plastic cups) or inaccurate estimation of the object's pose, which is due to improper processing of the depth data rather than misidentification.

6 Conclusion and Future Work

This paper presents a top-down object identification method for robotic picking in a cluttered environment. The method combines the use of RGB-D image data and historical identification performance to correctly identify a target item for robotic picking from a set of 25 commercially available items. Our approach is able to correctly recognize multiple instances of objects in arbitrary poses with high reliability and in a sufficiently fast time frame.

The system uses segmented images to identify a target item through the use of a hue-saturation histogram, geometric contour, visual BRISK features, and the estimated dimensions of the object in millimeters. A probabilistic model is generated offline from a library of global features generated by extracting above mentioned features from images covering a range of possible viewpoints of each object.

The system is shown to be able to correctly identify the target item from shelf bins containing either only the item or containing several other items (including another copy of the same item) and translate the identified position into real-world coordinates for a robotic manipulator to grasp. Our robotic picker was demonstrated at the Amazon Picking Challenge (APC) held at ICRA 2015 in Seattle, Washington. Outside the scope of the competition, the strength (and also its weakness) of our proposed approach is in the training dataset. However, the same approach could be applied to many different object identification problems (not just on e-commerce shelf picking) as long as the proper training dataset is used. Thus, a main challenge of our proposed approach is in scaling up as the training data needed is at least directly proportional to the number of perspective from which to identify the object.

One consideration for future improvement would be the use of high level indexing primitives such as generalized cylinders, geons, and superquadrics. This is in consideration of the system's use in commercial applications where the database contains thousands of items. The view or appearance-based approaches used here perform well on the 25 item dataset, but will likely be less optimized for higher number of items.

References

1. Andreopoulos, A., Tsotsos, J.K.: 50 Years of object recognition: directions forward. Comput. Vis. Image Underst. **117**(8), 827–891 (2013)
2. Prasad, D.K.: Survey of the problem of object detection in real images. Int. J. Image Process. (IJIP) **6**(6), 441 (2012)

3. Lowe, D.G.: Object recognition from local scale-invariant features. In: Proceedings of the 7th International Conference on Computer Vision, vol. 2, pp. 1150–1157 (1999)
4. Belongie, S., Malik, J., Puzicha, J.: Shape matching and object recognition using shape contexts. IEEE Trans. Pattern Anal. Mach. Intell. **24**(4), 509–522 (2002)
5. Richtsfeld, A., Mörwald, T., Prankl, J., Zillich, M., Vincze, M.: Learning of perceptual grouping for object segmentation on RGB-D data. J. Vis. Commun. Image Represent. **25**(1), 64–73 (2014)
6. Berner, A., Li, J., Holz, D., Stickler, J., Behnke, S., Klein, R.: Combining contour and shape primitives for object detection and pose estimation of prefabricated parts. In: Proceedings of IEEE International Conference on Image Processing (ICIP), pp. 3326–3330 (2013)
7. Marton, Z.C., Seidel, F., Balint-Benczedi, F., Beetz, M.: Ensembles of strong learners for multi-cue classification. Pattern Recogn. Lett. **34**(7), 754–761 (2013)
8. Bay, H., Tuytelaars, T., Van Gool, L.: Surf: Speeded up robust features. In: European Conference on Computer Vision, pp. 404–417. Springer, Berlin (2006)
9. Rosten, E., Drummond, T.: Machine learning for high-speed corner detection. In: European Conference on Computer Vision, vol. 1, pp. 430–443 (2006)
10. Leutenegger, S., Chli, M., Siegwart, R.Y.: BRISK: Binary robust invariant scalable keypoints. In: IEEE International Conference on Computer Vision (ICCV), pp. 2548–2555 (2011)
11. Rublee, E., Rabaud, V., Konolige, K., Bradski, G.: ORB: an efficient alternative to SIFT or SURF. In: IEEE International Conference on Computer Vision (ICCV), pp. 2564–2571 (2011)
12. Wang, L., Shi, S., Song, G., Shen, I.F.: Object detection combining recognition and segmentation. In: Computer Vision-ACCV 2007, pp. 189–199. Springer, Berlin (2007)
13. Amazon Picking Challenge [online] (2015). http://amazonpickingchallenge.org/index.shtml. Accessed 14 Sept 2015

Team KTH's Picking Solution for the Amazon Picking Challenge 2016

Diogo Almeida, Rares Ambrus, Sergio Caccamo, Xi Chen, Silvia Cruciani, João F. Pinto B. De Carvalho, Joshua A. Haustein, Alejandro Marzinotto, Francisco E. Viña B., Yiannis Karayiannidis, Petter Ögren, Patric Jensfelt, and Danica Kragic

1 Introduction

Throughout the last decades, automation has dramatically increased the efficiency of warehouses: complex automated conveyor belt and advanced vision and robotic systems are the norm in today's industry. Nevertheless, state-of-the-art robotic grasping and vision algorithms are not yet mature enough to cope with the vast variety of products, thus forcing many warehouse processes, such as the final packaging and handling of consumer goods, to be carried out by humans. For this reason, Amazon created the Amazon Picking Challenge (APC, recently renamed Amazon Robotics Challenge ARC), which invites competitors to engage in a contest for solving a simplified version of the general robot bin picking problem that considers only a limited set of objects.

The APC 2016 setup consisted of a shelf with 12 bins and a red tote, in which items can be stored as shown in Fig. 1. The participating teams were required to solve two different tasks:

- **Picking:** Pick a selection of objects from a shelf and place them into a tote. The jury specified which items were to be picked by the robot and from which bin of the shelf.
- **Stowing:** Execute the reverse process, i.e., pick objects from a tote and put them in different shelf bins. The goal was to empty the contents of the tote.

D. Almeida (✉) · R. Ambrus · S. Caccamo · X. Chen · S. Cruciani · J. F. P. B. De Carvalho · J. A. Haustein · A. Marzinotto · F. E. Viña B. · Y. Karayiannidis · P. Ögren · P. Jensfelt · D. Kragic
Robotics, Perception and Learning (RPL) Lab, School of Electrical Engineering and Computer Science, KTH Royal Institute of Technology, Stockholm, Sweden
e-mail: diogoa@kth.se; raambrus@kth.se; caccamo@kth.se; xi8@kth.se; cruciani@kth.se; jfpbdc@kth.se; haustein@kth.se; fevb@kth.se; yiankar@kth.se; petter@kth.se; patric@kth.se; dani@kth.se

A. Causo et al. (eds.), *Advances on Robotic Item Picking*,
https://doi.org/10.1007/978-3-030-35679-8_5

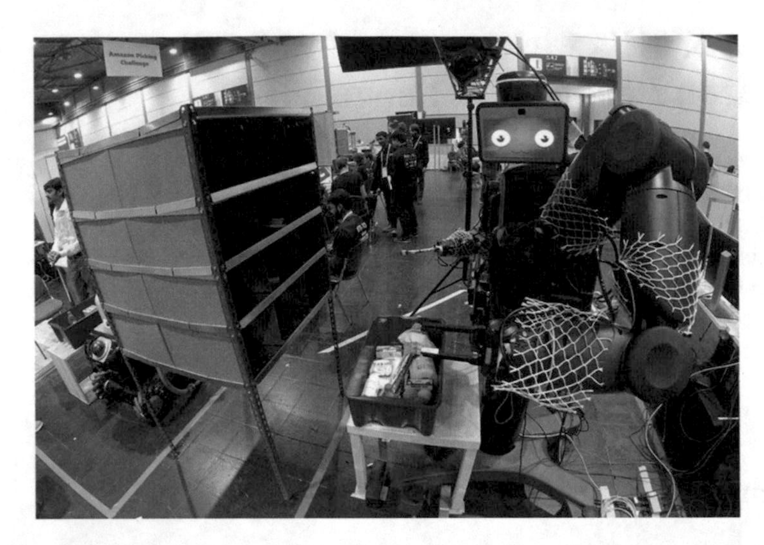

Fig. 1 Team KTH's Baxter robot at the Amazon Picking Challenge 2016 in Leipzig, Germany

Points were scored according to the number of objects successfully picked/stowed. Furthermore, picking incorrect objects or damaging them incurred a penalty. At the beginning of each trial, the organizers would slightly shift the position of the shelf in order to prevent teams to rely on pre-calibrated robot setups. In contrast with the APC 2015, the number of objects increased from 24 to 39.

Team KTH participated in the APC 2016 with the Baxter robot shown in Fig. 1. Baxter is a dual arm robot with 7 DOF per arm, which can be equipped with either suction gripping devices or parallel jaw grippers. We placed the tote between the robot and the shelf for better reach. Furthermore, we fixed the robot distance from the shelf to allow free movements of the two arms, avoiding hampering the robot manipulability. We complemented Baxter's hardware with three Kinect v2 cameras and two IDS camera. One Kinect was attached to Baxter's lower waist, to detect objects in the lower bins of the shelf and estimate the pose of the shelf itself. The other two Kinects were attached on a frame above Baxter, one facing down towards the tote and one facing towards the upper bins of the shelf. The two IDS cameras were placed on the wrists of the robot for close-up looks at specific shelf bins or at the tote. We also attached extensions to the suction cups of Baxter in order to improve the reachability of the robot inside the shelf.

In the following sections, we first provide an overview of our system and its high-level execution. Then, we describe our perception and manipulation pipelines and discuss some of the limitations of our system and potential directions for future improvements.

2 System Overview and High-Level Logic

We implemented the high-level logic of our system through a behavior tree (BT) framework [6] as shown in Fig. 2. The BT coordinates our perception and manipulation components. The perception and manipulation pipelines themselves are subdivided into smaller components (actions) which are implemented through the ROS actionlib protocol.[1] Action goals are managed by the BT, enabling independent development of the logic and actions. Two different strategies were developed for the *picking* and *stowing* tasks. In both strategies we prioritize certain perception methods to attempt faster detection of objects in the task description, and only use more robust but slower techniques if these methods failed.

2.1 *Stowing Strategy*

The stowing strategy implements three layers of perception methods to detect objects from the task description. The system first attempts to detect objects using the Kinect camera that is positioned looking straight down at the tote. If this method does not succeed, the BT commands the robot arms to move to different positions relative to the tote in order to detect objects using the wrist mounted cameras.

Once an object is detected, the system chooses a target bin in the shelf for its placement. This decision is based on both the object type (bigger objects are preassigned to the biggest shelf bins) and on the gripper type (some objects can be

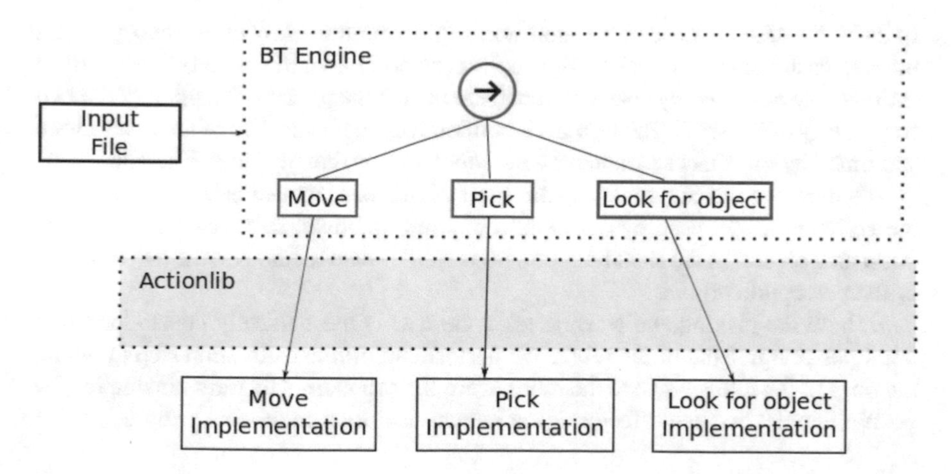

Fig. 2 Schematic overview of the BT implementation

[1]http://wiki.ros.org/actionlib.

grasped via suction, others require the parallel gripper). Since a single arm cannot reach every bin of the shelf, the leftmost shelf column is reserved for objects that can only be picked with the left arm, and the rightmost column for objects that can only be picked with the right arm.

2.2 Picking Strategy

While picking, the target objects are distributed over multiple shelf bins. A similar three layered strategy for detection is chosen, with the bin being defined by the task, instead of chosen by the system. Here, the priority is given to objects that can be picked with higher confidence.

3 Perception

The perception component of the system is responsible for estimating the pose of the shelf as well as detecting and estimating the pose of objects inside the bins and tote.

3.1 Shelf Pose Estimation and Tote Calibration

In the APC 2016 the pose of the shelf was slightly perturbed by the event organizers prior to each competition trial. We use the a priori known 3D model of the shelf to estimate the new pose by detecting the distance and the planar orientation of the two frontal legs of the shelf. The legs are identified from a planar slice of the point-cloud generated by the Kinect mounted on the robot's chest, cut at a proper height.

We then use the 3D model of the shelf along with its estimated pose to build the collision model used by our motion planning module, select camera poses for observing objects in the shelf bins, and filter RGBD data to improve the performance of the perception process.

In both the picking and stowing task, the tote is mechanically constrained to a fixed position in front of the robot. We perform an offline calibration step to obtain the pose of the tote relative to the robot using the robot arms forward kinematics, by positioning the two end-effector tips at known positions on the tote's edges.

(a) (b)

Fig. 3 Classification of APC'16 objects in our perception pipeline. (**a**) Example of textured objects. (**b**) Example of untextured objects

3.2 Object Detection

The object detection relies on three different perception strategies depending on the kind of object to be recognized and the camera to be used. The objects are divided into two main categories: textured and untextured. Figure 3 shows examples of these categories.

For the textured objects, the perception system relies mostly on *simtrack* [7] and a modified version of it, named *simtect*, tailored to maximize the chances of detection in a close look image.

Both of these methods rely on computing the pose of an object in an image using a set of 3D-2D correspondences. A model database is first constructed offline: each object mesh is rendered from a number of predefined orientations, and SIFT [10] keypoints, along with the corresponding depth, are accumulated in a 3D data structure. At runtime, for each camera image, SIFT keypoints are extracted and matched against the respective object 3D data structure. The pose of the object is then computed from the 3D-2D correspondences through a Levenberg–Marquardt optimization step [5].

The main difference between *simtrack* and *simtect* lies in the way correspondences are computed between a camera frame and the object database. While *simtrack* matches against the complete 3D data structure of a specific object (consisting of SIFT keypoints accumulated from all the mesh renderings), *simtect* matches the SIFT keypoints from each mesh rendering individually. This results in a more expensive method computationally, but increases the detection capabilities of the system, especially for objects with very similar textures on different sides.

For detecting untextured objects, the system relies on a different component that exploits the colored point-cloud from the Kinect cameras. After an initial noise filtering stage, it performs an RGBD Euclidian clustering on the filtered input pointcloud. It then analyses the color and the shape features on the isolated clusters to identify and classify the target objects [8].

The perception component provides the position and orientation of a selected target object and whether it is detected or not. To interface with the high-level logic, this perception component has been designed as an actionlib server, which accepts requests from an actionlib client. An action request carries information about the object to be identified, which allows to select the proper 3D model used by *simtrack* and *simtect*, or the object features map used by the RGBD server. Moreover, the request contains information on which shelf bin to search in and which cameras to use. Finally, information about objects in the same bin is also provided, to avoid detection conflicts and identify occlusions.

Once the perception server accepts a request, it redirects the message to the most appropriate sub-server to detect the target objects. The first sub-server is called *simtect server*, which uses *simtect* to detect textured objects. Since *simtect* is slower than *simtrack* but has a higher chance of detection, it is mainly used with the IDS cameras for closer looks at the bins or at the tote. The second sub-server is called *texture server*, which uses *simtrack* to detect textured objects; since *simtrack* is less computationally demanding than *simtect*, it is used in an initial step for a quick scan of the whole shelf using the images from the Kinect cameras. The last sub-server is called *RGBD server*, and it uses the colored point-cloud from the Kinect2 sensors to detect untextured objects. The overall structure of the perception software is depicted in Fig. 4.

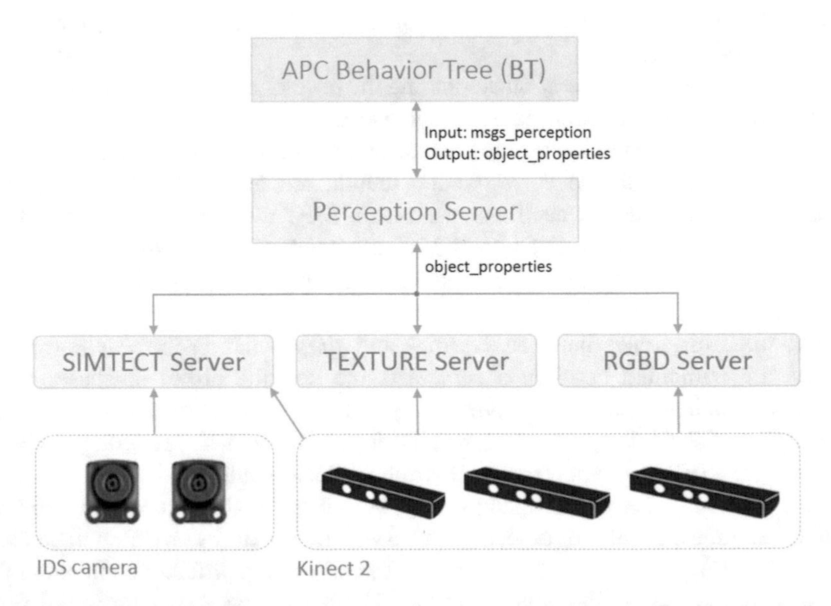

Fig. 4 The perception servers and their sensor interfaces; a request from the high-level planner is processed and, according to the object properties and the selected sensor, it is forwarded to the proper perception sub-server

4 Motion Planning, Pick and Place

The high-level logic system interacts with the robot platform through three different parameterized manipulation primitives, *pick*, *place*, and *move*. These manipulation primitives build upon the same motion planning procedures, which are detailed in the following sections.

While the Baxter robot is a dual arm system, we utilize only one arm at a time, allowing us to treat each arm as an independent system. While one arm is employed, the other arm is kept in a resting position. The decision on which arm to employ is made by the high-level logic and depends on the location of the target object as well as the object's type. For the picking task, both arms use suction cups as these maximize the chance of successful picking from inside the shelf. In contrast, for the stowing task we equip only one arm with a suction cup and the other with a parallel jaw gripper. This enables us to also pick objects from the tote that are not pickable with Baxter's suction cups.

4.1 Motion Planning

We distinguish between arm motions performed outside and inside the shelf. While the exact pose of the shelf is not known beforehand, it is guaranteed to lie within some bounded workspace region relative to the robot. Hence, motions outside this region are performed in a static and a priori known environment, whereas motions close to or inside it need to be adapted online to the actual shelf pose.

4.1.1 Outside-Shelf Motion Planning

For motion generation outside the shelf, we rely on a roadmap [4] computed offline. The nodes of this roadmap are manually defined arm configurations that either place the end-effector in front of the different shelf bins, on top of the tote, or in a resting position. For each shelf bin we select arm configurations that achieve high manipulability [11], thus allowing the robot to reach a large variety of desired end-effector poses inside or close to the bin without moving the arm into significantly different configurations. The edges of the roadmap are computed offline using the RRT* algorithm [3] provided by MoveIt [9].

An online motion planning query first connects the start and goal configuration to the roadmap. For this, it utilizes a local planner that first attempts to establish a connection based on linear interpolation in configuration space. If this fails due to collisions, the RRT* algorithm is used as fallback. If the start and goal configuration are successfully connected, these new nodes are permanently added to the roadmap. Thereafter, we apply the A* graph search algorithm [2] to search for a shortest path in the roadmap. The cost of a roadmap edge is the configuration space distance covered by the associated path.

4.1.2 Inside-Shelf Motion Planning

For motion generation inside the shelf, we rely on a heuristical approach. Rather than applying a sampling-based motion planner, we restrict motions to Cartesian straight line movements of the end-effector. While this limits the robot's ability to navigate inside the bins, it guarantees a predictable behavior. Before executing any such straight line movement, we perform collision checks between the shelf and the robot to avoid executing motions that collide with the shelf. Potential collisions with any bin content are ignored.

4.2 Pick and Place

Since the pose of a target object is not known beforehand, the system is required to select a picking pose online when an estimate of the target's pose is available. For this, our system evaluates a set of potential picking poses to determine a pose that is approachable without collisions. To approach a picking pose, we combine multiple of the aforementioned straight line movements, as illustrated in Fig. 5 for the case of picking from the top.

The set of potential picking poses is object dependent. We choose to encode this information for each object individually in the form of task space regions (TSRs) [1]. Here, a TSR describes a continuous set of end-effector poses relative to an object's frame, from which successful picking is feasible. In case the object is to be picked by suction, the TSRs describe the parts of the object surface to place the suction cup on. In case the object is to be grasped with the gripper, the TSRs describe grasping poses. In both cases, the system randomly samples the TSRs for picking poses and

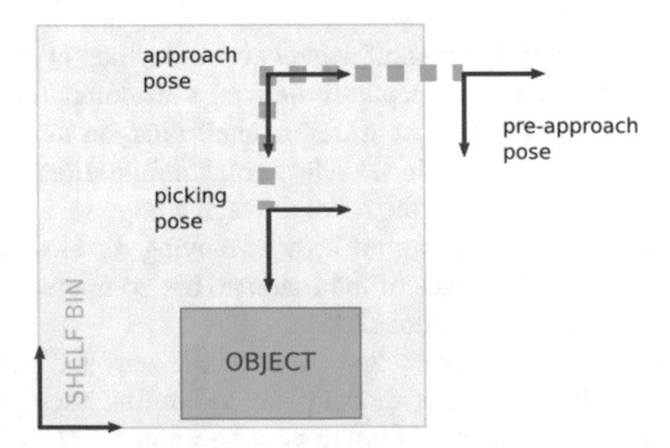

Fig. 5 An illustration of the picking approach from the top. For a given picking pose on the object surface, an approach pose and a pre-approach pose are computed. The approach pose is computed such that a straight line movement along the end-effector's z-axis moves it to the picking pose. The pre-approach pose is computed such that a straight line movement along the reverted x-axis of the bin frame moves the end-effector to the approach pose

evaluates their feasibility for execution. If an approachable picking pose is found, the system attempts to approach it and pick the object. Finally, the picking success is determined based on feedback provided by the suction cups or gripper closure.

The placement of objects inside shelf bins or the tote is performed in a similar manner. In both cases the end-effector is moved to a manually defined pose relative to the target bin/tote where the object is released.

5 Conclusions

This chapter presented the integrated picking system of team KTH for the APC 2016. We used a Baxter dual arm robot with some hardware enhancements in order to better suit the task.

The main limitations of our perception system was poor detection performance under heavy occlusions or poor lighting conditions, e.g., when the objects were located near the back of the shelf bins. Similarly, our manipulation system struggled with finding collision-free motion plans inside the shelf bins. We therefore propose the following directions for further improvement of our system:

- **Kinematics and motion planning.** One of the keys to a successful picking robot is a kinematic structure which is optimized for the task, since this, in turn, minimizes the workload on the motion planners. Baxter's bulky arms and its kinematic configuration, which is mainly designed for picking objects vertically from the top, made it difficult for our motion planners to find collision-free paths inside the bins.
- **Perception.** In order to improve the object detection rate of our perception system, state-of-the-art deep learning methods would be a promising alternative, as these proved robust to lighting conditions for many teams at the APC 2016.

References

1. Berenson, D., Srinivasa, S., Kuffner, J.: Task space regions: a framework for pose-constrained manipulation planning. Int. J. Robot. Res. **30**(12), 1435–1460 (2011). https://doi.org/10.1177/0278364910396389
2. Hart, P.E., Nilsson, N.J., Raphael, B.: A formal basis for the heuristic determination of minimum cost paths. IEEE Trans. Syst. Sci. Cybernet. **4**(2), 100–107 (1968)
3. Karaman, S., Frazzoli, E.: Sampling-based algorithms for optimal motion planning. Int. J. Robot. Res. **30**(7), 846–894 (2011). https://doi.org/10.1177/0278364911406761
4. Latombe, J.C.: Robot Motion Planning. Springer Science & Business Media, Berlin/Heidelberg (1991)
5. Marquardt, D.W.: An algorithm for least-squares estimation of nonlinear parameters. J. Soc. Ind. Appl. Math. **11**(2), 431–441 (1963)
6. Marzinotto, A., Colledanchise, M., Smith, C., Ögren, P.: Towards a unified behavior trees framework for robot control. In: Proceedings - IEEE International Conference on Robotics and Automation (ICRA), pp. 5420–5427 (2014). https://doi.org/10.1109/ICRA.2014.6907656

7. Pauwels, K., Kragic, D.: Simtrack: A simulation-based framework for scalable real-time object pose detection and tracking. In: Proceedings - IEEE/RSJ International Conference on Intelligent Robots and Systems (IROS), pp. 1300–1307 (2015). https://doi.org/10.1109/IROS.2015.7353536
8. Rusu, R.B.: Semantic 3D object maps for everyday manipulation in human living environments. KI - Künstliche Intelligenz **24**(4), 345–348 (2010). https://doi.org/10.1007/s13218-010-0059-6
9. Sucan, I.A., Chitta, S.: Moveit! (2017). http://moveit.ros.org
10. Wu, C.: SiftGPU: a GPU implementation of David Lowe's scale invariant feature transform (SIFT) (2007). http://cs.unc.edu/ãĂœccwu/siftgpu
11. Yoshikawa, T.: Foundations of Robotics: Analysis and Control. MIT Press, Cambridge (1990)

End-to-End Learning of Object Grasp Poses in the Amazon Robotics Challenge

Eiichi Matsumoto, Masaki Saito, Ayaka Kume, and Jethro Tan

1 Introduction

The Amazon Robotics Challenge (ARC)—formerly known as Amazon Picking Challenge— is an annual challenge issued by Amazon that is increasingly difficult. In this challenge, teams from both academia and the industry can enter to create the ultimate robotic system to solve what is considered by Amazon the final engineering challenges found in warehouse automation. Many decisions have to be taken on what approach to take for each subproblem when building such a robotic system for this task, which involves trade-offs and compromises given the limited amount of time and the resources each team has after receiving the detailed task specification. The "pick task" finals this year was thrilling as its winner was chosen by watching replays of their runs for the fastest cycle time after a tied best score between Team Delft and our team (Team PFN).

In this article, we briefly present our robotic system used in ARC 2016 that ended up placing second on the "pick task" and fourth on the "stow task." In particular, we would like to highlight our vision subsystem, which includes the main contribution of this paper: *end-to-end learning of object grasp poses*. We leverage convolutional neural networks (CNN) to not only provide semantic segmentations for each object, but also predict object grasp poses. This allows us to avoid the use of computationally intensive 6D object pose estimation by, e.g., point cloud registration and speed up the development process of our system.

Eiichi Matsumoto and Masaki Saito contributed equally to the work.

E. Matsumoto · M. Saito · A. Kume · J. Tan (✉)
Preferred Networks, Inc., Tokyo, Japan
e-mail: matsumoto@preferred.jp; msaito@preferred.jp; kume@preferred.jp; jettan@preferred.jp

A. Causo et al. (eds.), *Advances on Robotic Item Picking*,
https://doi.org/10.1007/978-3-030-35679-8_6

More information, source code used to run our system including scripts to train the vision subsystem as well as detailed models and images of the objects present in ARC 2016 can be found via [9].

2 Related Work

2.1 Amazon Robotics Challenge

In ARC, team entries are evaluated which includes approaches to object grasping in an unstructured environment. The latest ARC victors Team Delft [5] made use of a region-based convolutional neural network (CNN) to provide bounding box proposals for each object before executing a global point cloud registration algorithm to estimate the 6D pose of the objects. Based on this pose estimation, grasp strategies are determined in combination with handcrafted heuristics. Zeng et al. [13] developed a self-supervised deep learning system used in Team MIT-Princeton's entry that takes advantage of multi-view RGB-D data of the objects to obtain an initial pose for the iterative closest point algorithm. In [11], Schwarz et al. present evaluation of Team NimbRo's system during ARC 2016, which incorporates a perception system that makes use of CNN for both object detection and segmentation. For an overview of ARC 2015, we refer to [1] in which Correl et al. collected lessons learned from all team entries during that year, and to [2] wherein Eppner et al. described Team RBO's winning entry.

2.2 Object Grasp Pose Synthesis

A deep neural network model with two stages to predict and score grasp configurations is proposed in [6]. Guo et al. [3] improve this work by doing this in an end-to-end manner, while Nguyen et al. use a semantic segmentation based model to output object affordances in [8]. However, these methods are only tested in simple environments where objects are neither occluded nor occluding each other. On the other hand, self-supervising methods which for picking in unstructured environments have been proposed in [7, 10]. The biggest disadvantage of these methods is the training time they need.

3 System Overview

3.1 System Setup

Our system,[1] shown in Fig. 1, consists of two stationary FANUC M-10 *i*A industrial robot arms, which were each equipped with different end-effectors to handle different objects. Like the previous winner of ARC, we observed that a vacuum gripper was capable of handling the majority of the items. Therefore, we equipped the left arm with a vacuum gripper connected to a commercially available off-the-shelf Hitachi CV-G1200 vacuum cleaner allowing us to handle 37 out of 39 object types. However, an additional grasp method was needed due to the mesh texture of the pencil cup and the weight of the dumbbell. Hence, we deployed our second arm with a pinch-gripper to deal with these items and achieved full type coverage on all the objects with our system.

To control our system and process data, we used a laptop equipped with an Nvidia GeForce GTX 870M GPU and an Intel i7 6700HQ CPU running Ubuntu 14.04 with ROS Indigo. Additionally, two PCs running Microsoft Windows were used to

Fig. 1 Team PFN's robotic system entry in the Amazon Robotics Challenge 2016. The right arm had a pinch-gripper, while the left arm was equipped with a vacuum gripper, of which the tip can be actuated to rotate to an angle of 0°, 45°, or 90°. Both grippers had a length of 60 cm with 40 cm reach and were equipped with an Intel Realsense SR300 RGB-D camera, a Nippon Signal FX-8 3D laser scanner, and a VL6180X proximity sensor

[1]A video showing our system in action is accessible via http://goo.gl/ZavxX1.

interface the system to the Intel Realsense cameras and the Nippon Signal FX-8 laser scanners on each end-effector. Furthermore, four Arduino Uno boards connected to the VL6180X proximity sensors, a Bosch BME 280 environmental sensor, and the valves and the relay controlling the vacuum were also present in our system. The proximity sensors, which provided redundancy to the cameras and the laser scanners, were used to perform calibration at the very beginning of both the picking and the stowing tasks to calculate the offset and orientation of the shelf relative to the system.

3.2 Process Flow

To process tasks in our system, we use the "sense-plan-act" paradigm in combination with a global task planner, which puts the system into a certain state. For motions, we have defined joint space positions for the robots to capture the scene of the tote and the bins. Free space motions from and to these positions, as well as Cartesian space motions are commanded from the laptop to the robot controllers, which are responsible for the motion planning, generation, and execution. Because of the many heuristics present in our system, we will not describe them in this paper and refer to our source code [9].

Depending on the current state of the system and the given task, i.e., pick or stow, our task planner decides a next target using object-specific heuristics, such as size or weight. Afterwards, the robot moves in front of the location containing the target item to capture images of the scene from multiple directions. These images are then forwarded to our CNN—described in Sect. 4, which in turn provides segmentations for all detected objects and a map with ranked grasp candidates (object type, grasp pose, surface normal, prediction score) in the scene for all images in the local (i.e., the bin or tote with the target item) coordinate system. Grasp candidates are filtered out from this map if either (1) the prediction score is below a threshold, (2) the surface normals of its surrounding pixels are not consistent, or (3) the robot position is not safe because of predicted collision with the shelf or tote. Based on these candidates, our task planner then decides to either (a) pick the target item, (b) request coordinates for a grasp candidate of an item occluding the target item to move the occluding item to another location, or (c) completely give up on picking objects from the current location. To detect whether a grasp of an object has succeeded, we utilize the environment sensor described in Sect. 3.1 to measure the pressure of the vacuum. If failure is detected, we retry the grasp up to four times using slightly different positions and orientations. An exception to this process is when the pencil cup or the dumbbell is selected as target item, for which their grasp positions are determined using the iterative closest point algorithm found in PCL (Point Cloud Library).

4 Vision Subsystem

The problem definition in ARC is noteworthy in how objects were known beforehand. This not only allowed exploitation of object-specific heuristics, but also simplified the object recognition subproblem.

4.1 Model Overview

Figure 2 shows the architecture of the neural network used in our vision subsystem. In this network, the input is a four-channel RGB-D image with resolution of 320×240 pixels, while the output consists of two results, one for the semantic segmentation indicating 40 object classes (39 items and the background), and one for a confidence map with grasp poses in the coordinate system of the robots. Although our network was inspired by an existing fully convolutional encoder–decoder network [12] that consists of a "encoder" network with convolutional layers and a "decoder" network with deconvolutional layers, several settings are different. Specifically, we did not use any max pooling and unpooling layers, but instead employed convolutional and deconvolutional layers with stride 2 and kernel size 4. All layers in the encoder have ReLU activation, while all layers in the decoder have Leaky ReLU as activation. As with the original encoder–decoder network, we apply the batch normalization layers after the convolutional layers.

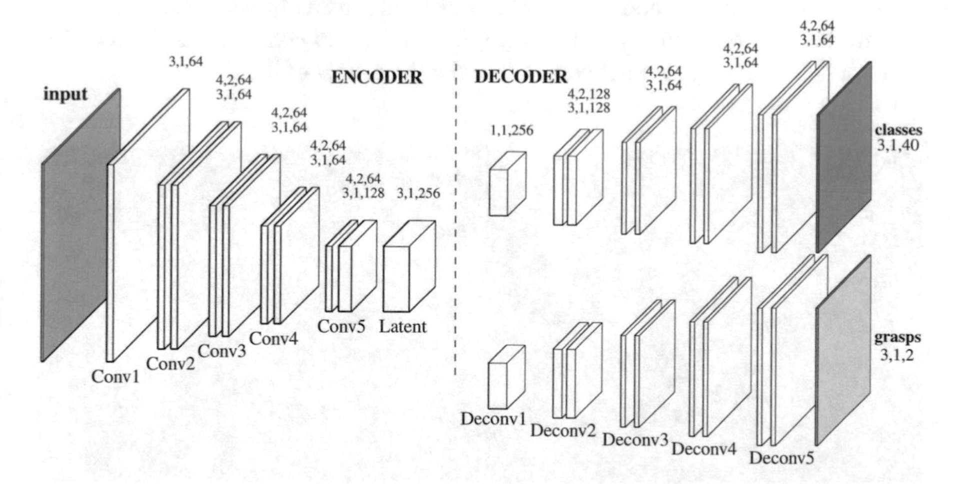

Fig. 2 Layer architecture of our CNN. The output of the encoder is connected to two identical decoders for outputting the semantic segmentation and graspability, respectively. The parameters in the convolutional and the deconvolutional layers are denoted as kernel size, stride, output channels

4.2 Data Collection

Using the robot arm, we have collected about 1500 images of all the bins and tote containing up to 12 items in a cluttered condition to simulate the competition scene. Although this number is relatively smaller than used in literature about CNN in general image recognition tasks, we found the output of the trained model to be workable for performing semantic segmentation on known objects. Moreover, we manually annotated the ground truth, i.e., per-object semantic segmentations and grasp positions ourselves with an in-house developed annotation tool to equalize the quality of annotation. Data collection and annotation took about 2 weeks in total.

4.3 Pre-training by CG Images

In order to improve the accuracy, we first pre-trained our model with an artificially generated CG dataset containing 100,000 images representing scenes from the tote. For the items in these images, we created 39 textured models with Blender, which allowed us to obtain the annotation of each CG scene. Despite these images being generated, we found that the quality of these images is comparable to images taken from the actual scene with real items, see Fig. 3. However, we empirically observed that the accuracy increase of the model is negligible, and therefore conclude that the performance gained by pre-training on CG images is not worth its effort. We conjecture that this is caused by the discrepancy of information in the depth channel between the CG images and images acquired from sensors in our system. We note that further analysis and experiments should be done to confirm this discrepancy and alleviate it by, e.g., adding noise to the depth channel of the CG images.

Fig. 3 Example of images in dataset used for training. (**a**) CG image with Blender models of items, and (**b**) image of an actual scene with real items

4.4 Training

For training, we initialized the weights in the encoder with HeNormal [4] and the decoder with zero centered normal distribution with $\sigma^2 - 0.02$. As optimizer, we used Adam with learning rate $\alpha = 0.001$. As loss function, we employed the sum of the two loss functions defined by $\mathscr{L} = \mathscr{L}_{\text{cls}} + \mathscr{L}_{\text{pos}}$, where $\mathscr{L}_{\text{cls}}$ denotes the softmax cross entropy loss representing the object class for each pixel, and $\mathscr{L}_{\text{pos}}$ denotes the binary cross entropy loss representing the grasp pose. During training, we observed that segmentation by the model trained with the default softmax cross entropy loss tends to ignore small objects such as the scissors. Therefore, we balanced the weight by multiplying it proportional to the reciprocal of the area: $\mathscr{L}_{\text{cls}} = \frac{N}{C} \sum A_i^{-1} \mathscr{L}_{\text{cls}}^i$, where N indicates the total number of pixels in the dataset, C the number of classes, A_i the number of pixels registered to class i in the dataset, and $\mathscr{L}_{\text{cls}}^i$ the softmax cross entropy loss for the class i.

Because of the inconsistency in image resolutions in our dataset, we resized all images to 320×240 pixels before performing random cropping to 224×224 pixels for training. The entire training process took 2 days on an Nvidia Titan X (Maxwell) GPU.

4.5 Inference

Similar to the training process, we resize retrieved RGB-D images to 320×240 pixels when inferring our network. Moreover, because the objects types in all locations are known, channels for object types that are not in a specific location are ignored.

5 Results

In preparation for ARC, we have evaluated the effect of the weight balancing described in Sect. 4.4 as well as using the depth channel in our network model. Results of our experiments can be found in Fig. 4. We observe that using depth information in addition to the RGB channels significantly improved the overall estimation accuracy of our model. Furthermore, our final model trained with weight balancing enabled outperformed our trained model without balancing in usefulness during the competition, because it contributed towards more reliable semantic segmentation for the small objects in ARC, albeit lowering the F-score for the large objects. The results of our final trained model are shown in Fig. 5, demonstrating accurate semantic segmentation of objects in dense, unstructured scenes. We notice that the grasp candidates are scored with consideration to the manipulation complexity of objects. For example, although the plastic coffee jar is

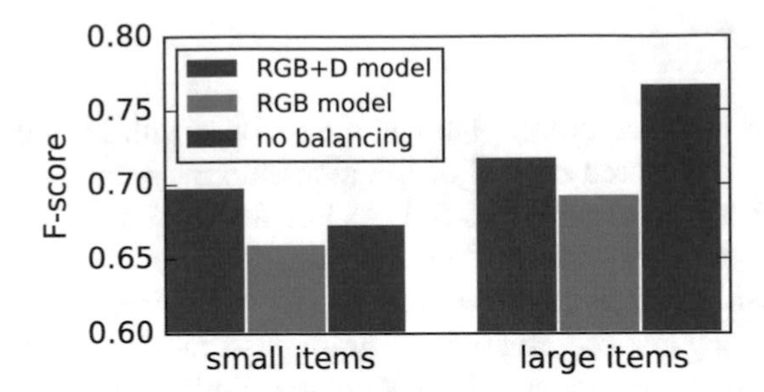

Fig. 4 Evaluation of semantic segmentation for variants of our model. The "RGB model" was trained without depth channel information, while the "no balancing model" did not use weight balancing for the object area size. Small items and large items are categorized according to an object's volume being lower or higher than 400 cm^3. F-scores are evaluated with a test dataset containing over 100 images in total

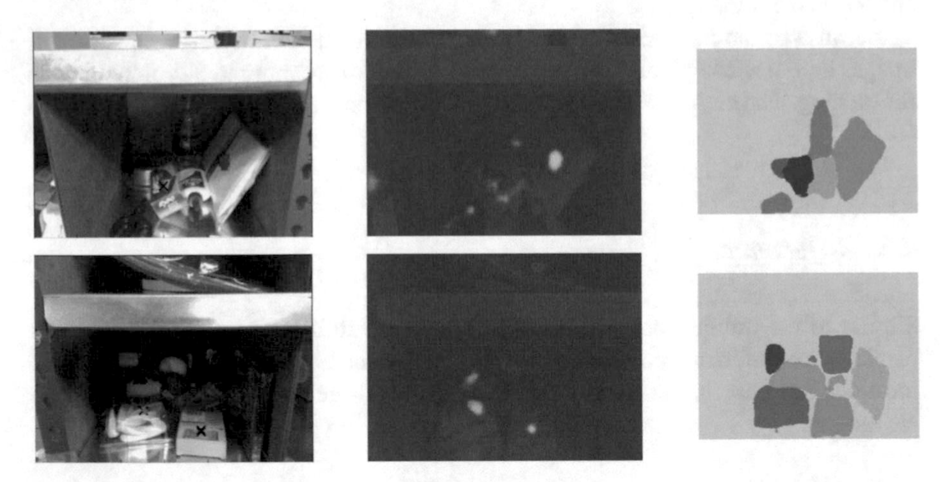

Fig. 5 Example of outputs produced by our network model containing (from left to right): (1) a color image with the selected grasp, (2) depth image with all available grasp candidates, and (3) semantic segmentation of objects in the scene

Table 1 Comparison of object grasp pose generation times

Team	Delft[a]	NimbRo picking[b]	MIT-Princeton	PFN
Time (s)	5–40	0.9–3.3	10–15	0.2

[a]Derived from [5] and analyzing the video linked in [5]
[b]Time is excluding ICP for 6D pose estimation

correctly detected in the semantic segmentation results of Fig. 5, no grasp candidates are proposed for this item in the grasp candidates map. Taking 200 ms to output these results, our vision subsystem was able to give us competitive advantage in data processing compared to the top scoring teams, as shown in Table 1.

6 Future Work

With the lessons learned during the competition, we note that many improvements can still be made to our system in order for it to be applicable in a real warehouse. In order to use our vision subsystem to its full extent, we have spent about 2 weeks to collect and annotate data. By applying techniques such as semi-supervised or weakly supervised learning, we can improve the scalability of our system to more object types as well as its robustness to unknown items. Furthermore, while our model was able to predict grasp poses for our vacuum gripper, we argue that a new method is needed for data collection and training in order for it to become applicable to more complex (e.g., higher DOF) end-effectors. Other functionalities requiring consideration include addition of visuomotor capabilities to the system in order to deal with complex object manipulation.

7 Conclusion

In this paper, we briefly described our system entry to ARC 2016 and proposed a vision subsystem using CNN to enable end-to-end learning to predict object grasp poses from RGB-D images. Thanks to the generalization abilities of deep learning models, our trained model was able to successfully produce semantic segmentation of objects in unstructured environments where items can be occluded, while also outputting grasp poses for these objects within 200 ms. The effectiveness of our method was demonstrated in the competition, where our entry placed among the top scoring teams.

Acknowledgements The authors would like to thank Tobias Pfeiffer, Taizan Yonetsuji, Yasunori Kamiya, Ryosuke Okuta, Keigo Kawaai, Daisuke Okanohara for their contribution and work done in the Amazon Robotics Challenge as part of Team PFN, and FANUC Corporation of Japan for providing the robotic arms and technical support.

References

1. Correll, N., et al.: Analysis and observations from the first Amazon Picking Challenge. IEEE Trans. Autom. Sci. Eng. **15**(1), 172–188 (2016)
2. Eppner, C., et al.: Lessons from the Amazon Picking Challenge: four aspects of building robotic systems. In: Proceedings of the Robotics: Science and Systems, University of Michigan, Ann Arbor, 18–22 June (2016)
3. Guo, D., et al.: Deep vision networks for real-time robotic grasp detection. Int. J. Adv. Robot. Syst. **14**(1), 1–8 (2016)
4. He, K., et al.: Delving deep into rectifiers: surpassing human-level performance on ImageNet classification. In: Proceedings of the International Conference on Computer Vision, CentroParque Convention Center, Santiago, 11–18 December (2015)

5. Hernandez, C., et al.: Team delft's robot winner of the Amazon Picking Challenge 2016. In: Behnke, S., Sheh, R., Sarıel, S., Lee, D. (eds.) RoboCup 2016: Robot World Cup XX. RoboCup 2016. Lecture Notes in Computer Science, vol. 9776. Springer, Cham (2017)
6. Lenz, I., et al.: Deep learning for detecting robotic grasps. Int. J. Robot. Res. **34**, 705–724 (2015)
7. Levine, S., et al.: Learning hand-eye coordination for robotic grasping with deep learning and large-scale data collection. Int. J. Robot. Res. **37**, 421–436 (2017)
8. Nguyen, A., et al.: Detecting object affordances with convolutional neural networks. In: Proceedings of the IEEE/RSJ International Conference on Intelligent Robots and Systems, Daejeon Convention Center, Daejeon, 9–14 October (2016)
9. Pfeiffer, T., et al.: Team PFN source code (2016). https://github.com/amazon-picking-challenge/team_pfn. Accessed 18 Aug 2016
10. Pinto, L., Gupta, A.: Supersizing self-supervision: learning to grasp from 50K tries and 700 robot hours. In: Proceedings of the IEEE International Conference on Robotics and Automation, Waterfront Congress Centre, Stockholm, 16–21 May (2016)
11. Schwarz, M., et al.: NimbRo picking: versatile part handling for warehouse automation. In: Proceedings of the IEEE International Conference on Robotics and Automation, Marina Bay Sands, Singapore, 29 May–3 June (2017)
12. Yang, J.: Object contour detection with a fully convolutional encoder–decoder network. In: Proceedings of the IEEE Conference on Computer Vision and Pattern Recognition, Caesars Palace, Las Vegas, 26 June–1 July (2016)
13. Zeng, A., et al.: Multi-view self-supervised deep learning for 6D pose estimation in the Amazon Picking Challenge. In: Proceedings of the IEEE International Conference on Robotics and Automation, Marina Bay Sands, Singapore, 29 May–3 June (2017)

A Systems Engineering Analysis of Robot Motion for Team Delft's APC Winner 2016

Carlos Hernandez Corbato and Mukunda Bharatheesha

1 Introduction

The Amazon Picking Challenge (APC) 2016 involved two manipulation tasks, to *pick* and *stow* products from an Amazon shelf, which addressed some of the challenges for reliable picking in a real warehouse, such as diversity of products, cluttered spaces, uncertain environment conditions, or full autonomy. Team Delft robot won both competitions with a solution based on 3D cameras for deep-learning based item detection, a planning-based solution for the motions of an industrial manipulator, a custom gripper, and the integration of off-the-self software and application-specific components with the Robot Operating System (ROS).

System-level integration is acknowledged as one of the key issues in the development of autonomous robotic systems [1, 4, 5, 8], as in other modern complex systems. Therefore, in this chapter we take a systems engineering stand to perform a postmortem analysis of Team Delft's solution for robot motion, understanding the benefits and limitations of the planning approach taken. We focus on the picking task for simplicity, because it includes all the challenges and proved to be more demanding,[1] although it is important to note that the designed solution solved both

[1]The structure of the problem resulting for the relative perpendicularity of gravity and the shelf's bins opening, together with the bins form factor with large depth/opening ratio proved still more of a challenge for robotic manipulation than the picking from the tote required for the stow task.

C. Hernandez Corbato (✉)
Delft University of Technology, Delft, The Netherlands
e-mail: c.h.corbato@tudelft.nl

M. Bharatheesha
Caspar AI B.V., Rotterdam, The Netherlands
e-mail: mukunda.bharatheesha@caspar.ai

A. Causo et al. (eds.), *Advances on Robotic Item Picking*,
https://doi.org/10.1007/978-3-030-35679-8_7

tasks. For a detailed analysis of the overall solution, we refer the reader to [4]. Our analysis framework is presented in Sect. 2. Section 3 analyzes the design of the motion subsystem, and its resulting properties and performance. Finally, Sect. 4 discusses general issues on the design of complex autonomous robots, looking into future challenges.

2 A Framework to Analyze an Autonomous Robot Design

The development of modern robotic systems, such as those in the APC, requires the *integration* of diverse technologies, e.g., 3D cameras, deep neural networks, planning, robot arms, and involves different disciplines, such as mechanical design, machine learning, control and software engineering. Model-based Systems Engineering provides methods and tools to consistently integrate the information, assumptions, and decisions through the development of the system, from high-level capabilities operational needs to detailed design decisions, and is thus an appropriate framework for the design and implementation of autonomous robots.

Besides the integration challenge, to address the capabilities needed, the selection of the control schemas at different levels of the solution architecture is based on assumptions about the structure of the task and the environment, and the associated uncertainty. For this reason, our framework also includes model of levels of robot automation [4]. This model categorizes the different control schemas, based on the degree to which they address uncertainty with prior assumptions or runtime data, and the resulting quality attributes (QAs) in the robot capabilities they address.

2.1 *Functional Analysis Under the ISE&PPOOA Method*

Autonomous robots are complex systems that require powerful systems engineering methods to allow teams[2] to build solutions that meet the demands of industry and society, especially QAs such as reliability and safety, and associated non-functional requirements (NFRs).

[2]It is important to note that we are not directly referring here to the robots participating in the ARC, for example. These are research prototypes whose development is only concerned with functional requirements and some non-functional requirements, such as speed or error recovery, but only to the extent that they directly relate to the scoring in the competition. Actually, to the knowledge of the authors, none of the teams in the ARC competition followed an MBSE approach. In the case of Team Delft, the leading author partially followed the ISE&PPOOA methodology during the earlier stages of requirement analysis and conceptual design. However, to apply any MBSE methodology for the complete development of the system, the team needs to be familiar with it, and this was not the case.

The ISE&PPOOA[3] process is an integrated systems and software engineering methodology that adopts the functional paradigm for model-based systems engineering. The systems engineering subprocess in ISE&PPOOA proceeds as follows:

1. Identify the operational scenarios, obtaining the needs, that are the main input for the
2a. specification of the capabilities and functional requirements of the system, and in parallel the
2b. specification of the relevant QAs and associated system NFRs. From this, the solution is created by iterative refining of
3. the functional architecture, which is obtained by transforming the previous level functional requirements into a functional hierarchy, and the
4. physical architecture, obtained by the allocation of functions to the building blocks of the solution, refined by design heuristics and patterns to address specific QAs.

In this chapter, we use the basic concepts in ISE&PPOOA to analyze the design of the motion subsystem in Team Delft's robot. The functional approach of ISE allows to abstract the architectural properties of the solution, independently of particular implementation details. Additionally, ISE&PPOOA explicitly incorporates QAs through the application of heuristics and design patterns to define the physical architecture of the solution. This helps understand how the assumptions and the design decisions impacted the architecture of the solution and the performance obtained in the capabilities required for the robotic challenge.

2.2 Levels of Robot Automation

The control architecture of an autonomous robot is designed to automate its runtime actions despite a certain level of uncertainty, by exploiting knowledge and assumptions about the structure of the task and the environment. The decisions made to select specific control schemas or patterns embed assumptions that critically determine the resulting quality attributes for the robot's capabilities (e.g., reliability of the grasp, speed of pick and place, etc.). We have elaborated a characterization of control schemas in *levels of robot automation* [4]. This characterization is aimed at supporting design decisions on *how* to automate each function in the system, i.e., which control pattern to use, considering the uncertainty level and the requirements on the QAs. Previous models, such as the one in [6], focused on *which* function to automate, and which ones to allocate to a human operator.

The base criteria to differentiate our levels of robot automation is time: when the information that decides the action is produced (at design-time or at runtime), and how it is used to generate the action (periodically at a certain frequency in a loop with runtime data, or in an open-loop manner based on strategies predefined at design-time).

[3] http://www.ppooa.com.es.

Level 0 corresponds to classic open-loop automation. A complete and static model of the task and the environment is assumed and no runtime data is used. This allows to specify a fixed sequence of motions and actions during development, which are blindly executed by the robot to perform the task.

Level 1 includes sensing information at runtime to perform a binary verification of the assumptions about the state of the system or its environment. This is the case of a sensor to detect a correct grasp of an object, or if it is dropped. Note that, no matter whether the sensory data comes a posteriori or prior to the action, the latter is specified at design-time, and only a discrete selection is performed at runtime. Level 1 mechanism overcomes some of the limitations of level 0, e.g., addressing uncertainty through basic error handling mechanisms. However, a complete discrete, predefined model of the environment and task is still assumed.

Level 2 corresponds to the so-called sense-plan-act paradigm. A prior model of the environment and the task is updated with sensory data at runtime to compute the control action. This level accounts for more complex environments, through perception capabilities that estimate the operational state an account for limited, static uncertainties, and planning to produce actions sequences adapted to the perceived situation. Still a perfect motor model is assumed to execute the planned actions, which results in well-known limitations.

Level 3 is the classic feedback control, in which the robot actions continuously adapt to the perceived state through closed-loop control. This level can cope with higher levels of uncertainty in the variables considered at development-time for the design of the control system, and can also address bounded unknown perturbances.

Level 4 considers more advanced control schemas including *prediction* to compute optimal control actions, considering not only current runtime sensory data and state estimation, but future states too. Even these control schemas still assume a prior model of the robotic system and its environment, though in this case it can be a complex model including dynamics.

Level 5[4] Control architectures in this level are no longer bounded by the models at design-time, but incorporate mechanisms to learn and improve upon those models using runtime data, for example, exploiting machine learning techniques, and adapt their control policies or learn new ones for new tasks, for example, with reinforcement learning.

It is important to note that the control architecture of a complex autonomous robot typically leverages control solutions for functions at different levels in the functional hierarchy. For each element in the architecture a solution is chosen from the appropriate robot automation level, based on a trade-off analysis between the QAs (e.g., reliability, impacted by the level of uncertainty), function performance, e.g., speed, and cost, e.g., in terms of resources and development effort, heavily impacted by the use of off-the-shelf. Following we examine the design decisions

[4]This level of robot automation has been included, compared to our first model in [4], to account for robotic systems that can adapt and learn.

in Team Delft's motion subsystem, using the framework provided by ISE&PPOOA and the levels of robot automation to analyze architected solution in terms of the control patterns selected and the resulting measurement for its QAs.

3 Motion Subsystem Design

Pick and place systems require a generic set of robotic capabilities, regardless of the specific application, namely: locate and identify objects, attach or grasp individual objects, and manipulate or move them [5]. In this section, we analyze Team Delft's solution for the motion capabilities following the functional approach presented in Sect. 2.1, which allows to abstract general (functional) design considerations from the specific solution adopted.

3.1 Motion Requirements

The main operational scenarios for the motion subsystem in Team Delft solution were: (1) move the camera attached to the robot's end-effector to obtain images of the bin for the next target item, and (2) plan and execute the motions to retrieve the target item from the bin and place it in the tote. The operational needs of these two scenarios were transformed into the detailed functional requirements, NFRs, and design constraints in Table 1 and assumptions in Table 2, some of which come from the design of the overall system, the competition rules and desired quality attributes in the system.

3.2 Robot Manipulator

A typical decision in manipulation applications is the selection of the robot. It is a core building block in the physical architecture of such systems that is allocated in the motion functionality.

A solution based on an off-the-shelf industrial manipulator and the MoveIt! library[5] was deemed the most feasible and desirable, given the resources available and also the skills in the team, very familiar with motion planning in industrial manipulators.[6]

[5]The documentation for the various APIs referred in this text can be accessed via: http://moveit.ros.org/code-api/.

[6]At an early stage a cartesian robot was discarded because it required extensive and specific mechanical and software development for its control. Its potential benefits in speed and simplicity of the kinematics were not as immediate in the shelf configuration of APC as in 2017 edition, when such a cartesian manipulator system won [5] by using that edition's possibility to have a custom horizontal storage system, which made the task decomposable into a series of straight motions by exploiting the cartesian structure in the environment.

Table 1 Requirements for the motion subsystem in Team Delft's robot

ID	Requirement
FReq.1.1[a, b]	The system shall move the camera in the robot end-effector to an appropriate pose in front of the target bin avoiding collisions
Related:	From ON.1 and QA harmless
FReq.2.1	The system shall generate planned motions to reach with the gripper all locations inside the shelf's bins that are relevant for grasping items
PReq.5.2	The robot shall execute the trajectory to reach a target location as fast as possible
Related:	Derived from speed req., related to QA reliability, as this maximizes the opportunities to pick items in the allotted time
FReq.5.1	The system shall find a collision free path to grasp the product, and a retreat path to retrieve the item from the shelf
FReq.3	The system shall achieve and hold a firm grasp on all different products
Cons. 1	The robot shall move following a velocity and acceleration profile that guarantees a reliable grasp when holding an item
Related:	From QAs reliability and harmless, to prevent dropping a grasped item when moving
Cons. 2	The robotic system shall fit inside an area of $2 \times 2\,\text{m}^2$ separated 20 cm from the front of the shelf and the start of the task
Related:	From the competition rules

These are more refined requirements based on the overall system analysis in [4]
[a]FR: functional requirement. NFR: non-functional requirement. Cons.: Constraint
[b]The numbering is consistent to the one reported in [4]

Table 2 Assumptions for the motion subsystem, some are requirements for other subsystems in Team Delft's robot

Assumption 1 (from Req.4)	A suitable grasp surface is always directly accessible from the bin opening that allows to grasp and retreat holding the product, and no complex manipulation inside the bin or the tote is needed. This way the 'level 2' assumption of environment invariance holds
Assumption 2	Not all the locations inside the bins are relevant for grasping. Due to gravity and based on the dimensions of the objects, no grasp candidates are expected in the higher part of the bins
Assumption 3	Environment out of the shelf is known and static
Assumption 4	Environment inside the bin is unknown and static during grasp operation. This required that the collision scene of the bin needs to be updated every time a grasp attempt is executed

A configuration with an off-the-shelf robot manipulator and an additional axis for enhanced reachability would render the required precision and speed. Multiple configurations were considered (Fig. 1a and b), constrained by the robot cell dimensions Cons. 2, and subsequently evaluated according to the requirement for reachability (FReq.3) using the MoveIt! workspace analysis tools [7].

The configuration shown in Fig. 1a was finally chosen, as the reachability analysis results ensured that the gripper could reach all bins with adequate maneu-

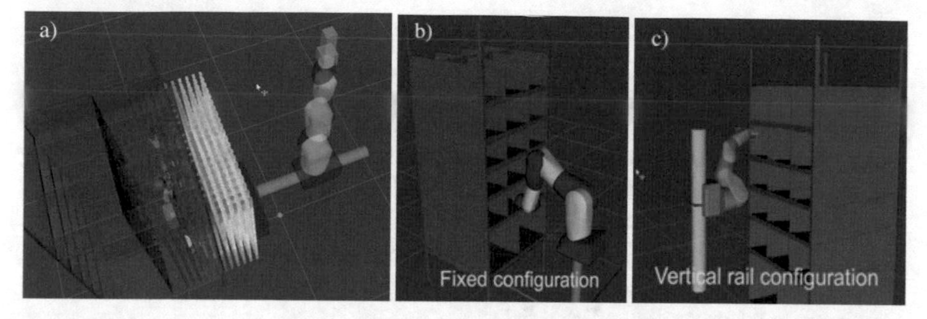

Fig. 1 (**a**) Reachability Analysis results of the selected robotic system configuration. The color of the marker in the figure indicates the number of IK solutions available to reach the corresponding cartesian location in the robot cell. Red markers are for locations that have very few IK solutions. (**b** and **c**) Other configurations analyzed

verability. An important aspect that can be observed in the figure is that some of the deep corners of the bins have no markers, indicating that those regions are not accessible with this system configuration. This limitation was compensated by slightly increasing the length of the suction tool (represented by the solid gray block in the robot's end-effector in Fig. 1a.

3.3 *Motion Software Module Design*

To fulfill the motion requirements of both the picking and stowing tasks, the motion module[7] was built on two fundamental motion primitives, namely, coarse motions and fine motions, that relied on the assumptions in Table 2. Coarse motions were offline generated trajectories between predefined start and goal positions in the robotic cell. Fine motions involved online (cartesian) path planning for performing object manipulation in the bins or the tote. In the following sections, these two primitives will be explained in further detail.

3.4 *Offline Coarse Motions*

Assumption 3 of a static environment allowed us to select an efficient 'level 0' solution for the robot motions outside of the shelf and the tote, using predefined joint

[7]The open source software repository hosting the code for the motion module and the rest of the ARC 2016 software is available at: https://github.com/warehouse-picking-automation-challenges/team_delft.

trajectories, or *coarse motions*, between configurations associated to key positions in the workspace.

These configurations, named Master Pose Descriptors, are robot joint states set at appropriate values in front of each bin of the shelf for the functions FReq.1.2 to take a bin image and FReq.5.1.1 to approach target object. Our choice of having the camera mounted on the manipulation tool entailed that we have two master pose descriptors per bin, namely, the *camera* and the *bin*[8] master poses.[9] Collision checking using 3D model (URDF) of the elements in the robot cell (enlarged to account for known uncertainties, such as 3 cm deviations in the positioning of the shelf) ensured safe motions (FReq.5.1).

Subsequently, a *trajectory database* with around 250 trajectories between different master poses was generated using the RRT-Connect randomized path planner via MoveIt!. Other planner options were analyzed, such as RRT* and the OMPL[10] implementation KPIECE, but RRT-Connect outperformed both of them for our planning problem (characterized by the robot kinematics chosen and the geometric structure of the shelf and tote). The main reason for RRT-Connect outperforming was that it maximally exploited the benefit of the linear joint (rail), because it explores solutions starting from both the start and end nodes, using Euclidean distance in configuration space.

3.5 Grasping and Manipulation: Fine Motions

From a motion perspective, the grasp strategy for all objects consisted of a combination of linear segments. We call these segments as *Approach*, *Contact*, *Lift*, and *Retreat*. The segment names are indicative of the motions that they are meant for. This solution maximally exploited the geometry of the bins, and Assumption 2.

The idea of fine motions is an adaptation of the approach in the standard pick and place pipeline of MoveIt!, where cartesian path planning is used during the pre-grasp approach and the post-grasp retreat stage of the pipeline. We removed the standard filter stage in the MoveIt! pick and place pipeline consisting on evaluation of the reachable and valid poses. To optimize the end-to-end time performance of the planning step in the overall application flow, this filtering was divided into multiple steps.

The first one is performed during the function grasp synthesis, where impossible grasps are eliminated heuristically, exploiting knowledge of the products and their graspability given their relative pose inside the bin. Then second filtering is performed by the motion subsystem, by checking the feasibility of the manipulation plan associated with the grasping candidate. The feasibility is determined by avoiding collisions, and obtaining a feasible joint trajectory (Inverse Kinematics solution).

[8]Starting configuration to approach target object.

[9]Similar master pose descriptors were also defined for the tote drop-off locations.

[10]MoveIt!, the framework chosen to implement our motion module, uses OMPL for planning.

3.5.1 Collision Avoidance

During grasping and posterior manipulation to retrieve the object from the bin, collisions need to be avoided to prevent damaging items or the shelf and to lose a grasped item. For collision checking the following information is used: runtime data, consisting of the occupancy octomap from the scanned point cloud, from which the points corresponding to the target object are removed to allow for contact in the case of suction grasping, and additional prior information, consisting of the 3D model of the shelf in URDF, and a collision model for the specific bin (mesh), which position is nevertheless computed online in the perception pipeline.

3.5.2 Generation of the Complete Grasp Plan

Departing from the pose of the target item, the grasp synthesizer module generates heuristically a set of key waypoints for the start and end of each linear segment. These waypoints are input to the motion module, which obtains the manipulation plan proceeding as follows:

1. The key waypoints are sequentially checked for collision and feasibility.[11] If any one of them is in collision, the corresponding grasp pose is discarded and an alternative one is requested.
2. Once all the key waypoints are collision free, each linear motion segment is computed using cartesian planning[12] with collision checking enabled. Similar action is taken if any of these segments are in collision.
3. If all the linear motion segments are collision free, a final planning in joint space is done, using RRT-Connect because of its fast solution times.[13] This is required because, the final joint configuration at the end of the Retreat segment resulting from the cartesian planning need not necessarily match the starting joint configuration in front of the bin or tote from where the coarse motions start. This is natural because of the redundancy in the system's degrees of freedom. The final planning ensures that such a mismatch does not exist which would otherwise lead to a motion safety violation.[14] To obtain last joint configuration in the cartesian trajectory, the inverse kinematics solver is configured to minimize the steady state error,[15] which yields the closes joint solution to the configuration goal, minimizing configuration changes.

[11]Using the `getPositionIK` service call.

[12] `computeCartesianPath` in the MoveGroup API.

[13]`plan` the MoveGroup API.

[14]A motion safety violation is triggered whenever the starting configuration of the robot does not match the starting configuration in the trajectory that is about to be executed.

[15]*Distance* setting in the Trac-IK solver used.

These steps ensure that all the desired motion segments for manipulating the object of interest are generated as required and are collision free. These segments were then stitched together before being executed on the robot, as explained in the following section.

3.6 Trajectory Execution

After obtaining a feasible grasp plan (complete joint trajectory to retrieve the target item), it had to be executed. To address requirements PReq.3.2 and Cons. 1, two solutions were designed that prepared the trajectory for execution[16]: trajectory stitching and I/O synchronization.

3.6.1 Trajectory Stitching

The motion stitching module accepts all the motion segments that are generated as explained in the previous section, plus the coarse motion trajectories from the corresponding bin to the tote drop-off location. The stitching process combines the joint state configurations from each segment into one single motion plan and time parameterizes[17] it so that it results in a executable trajectory for the robot.

With respect to PReq3.2 on the end-to-end speed of the robot motions, by stitching multiple motion segments we get rid of overheads such as goal tolerance checks at the end of each segment execution. This indeed provided us quite a significant time gain while executing the motions. In relation to Cons. 1 to prevent dropping items, the stitching performed object specific velocity scaling for the trajectory to adapt the motion to the product, for instance, moving at low speeds when carrying heavy objects such as the dumbbell, socks, and kitchen paper roll.

3.6.2 Input/Output Synchronization

The final and a critical component of the motion module is the I/O handling, designed to ensure the end-effector (suction or pinch) is actuated at the right times along the trajectory, to guarantee successful grasps in relation to Cons. 1. The timely activation of the suction gripper was critical. For instance, the vacuum pump needed a couple of seconds before full suction power was realized, which meant it had to be switched on before the suction cup reached the contact point. However, turning the suction on too early could also pose problems. For example, the items with a loose plastic covering could get suctioned before the contact point was reaching,

[16]The time parameterized trajectory was finally executed in the robot controller through the MoveGroup API, `execute`.

[17]`computeTimeStamps` from the TrajectoryProcessing API was used for this purpose.

leading to either an unstable or a failed grasp (Req.3). In order to address this, a custom trajectory tracking module[18] was developed that provided continuous joint state information.

The trajectory tracking module used an event based approach. The events correspond to reaching the key waypoints along the trajectory, and the tracking module used a gradient descent based approach based on the distance to those points (level 3 based on runtime data from the encoders). These events were used not only to trigger the suction, but also to trigger the pressure sensor feedback, for example, to evaluate the success or failure of a grasp at the end of the retreat segment. This runtime data allows to cancel the open-loop execution of the trajectory if an error is detected at certain points, and take appropriate actions (level 1 mechanism), for example, re-try the grasp if the pressure sensor detects no vacuum seal after retreat (i.e., the item is not attached).[19]

4 Discussion and Concluding Remarks

Team Delft's robot was based on an overall sense-plan-act (level 2) solution, that relied on detailed solutions for specific motion functions that ranged from level 0 to level 3, as discussed along Sect. 3 and summarized in Table 3.

It is important to note that intuitively, neglecting uncertainty, performance and simplicity (a desired property when building a system) are better for lower levels of robot automation. Although they are brittle to uncertainty, in a partially structured and known environment such as the one in APC16, simple solutions error-detection and handling at level 1 seem sufficient. However, our planning and collision avoidance solution is based on the assumptions that an unobstructed grasp is directly accessible for the target item and the environment is static during the operation, which simplifies the manipulation problem. This does not scale in general for more complex manipulation scenarios, for example, inside a densely packed bin. Reactive

Table 3 Levels of robot automation of the different elements in the design of Team Delft's motion module

Level	Design element
Level 0	Coarse motions (predefined) to move the camera to take images and to move the picked object from the shelf to the tote
Level 1	Velocity scaling of the trajectories according to the product grasped, during the stitching process
Level 1	Detect suction grasp failure using a pressure sensor in the nozzle
Level 2	Fine motions (online cartesian planning) to grasp and retrieve the items
Level 3	Trajectory tracking to synchronize actions along the trajectory

[18]Based on the `/joint_states` topic.

[19]More details on this finite state machine approach to error handling can be found in [4].

grasping that allows and even exploits contact (level 3) is more general and robust approach. To account for handling novel objects, level 5 methods are needed, such as deep learning, to recognize them and adapt the control actions, as reported by successful entries in 2017 edition.

The functional-level analysis provided by the ISE&PPOOA methodology allows to identify useful patterns that we intuitively applied to design our solution. For example, we use a service-based blocking architecture for the motion module, which renders a deterministic behavior, e.g., preventing race conditions. This is a well-known pattern for safety and reliability [2]. However, this impacted negatively on the complexity of the implementation tracking module, which required a dedicated thread of execution. An asynchronous *ROS action* based architecture is more scalable for systems that require level 3 closed-loop solutions (multiple control threads).

In relation to the synchronization of actions along the trajectories, we must note that MoveIt! allows for I/O handling to be synchronized with trajectory execution by incorporating the external devices as joints in the planning. However, this poses some problems for binary joints like our suction pump. A decision was made to create our trajectory tracking module, which creates a more reusable solution, independent of MoveIt!, that also allows to synchronize sensors in addition to actuators (modularity and integration vs. reuse an existing solution).

Other design strategies were intuitively applied to the functional architecture of our solution. The function to generate grasp candidate(s) relied on heuristics based on the product type and the current pose in the planning scene, as well as a looped interaction with the function to generate the associated waypoints, which relies on the same information. Therefore, both functions were allocated to the same module, the grasp synthesizer, in an example of the systems engineering strategy that advocates for jointly allocate functions that are closely related. This decision primed the benefits of integration, in terms of simplicity and efficiency, over the re-usability of a more modular design.

The functional architecting of our solution also presented some oversights in our functional design resulting in some limitations. For example, in our functional design we did not considered explicitly the function "move item," and the associated requirement was implicitly allocated to the gripper and the suction system. This neglect resulted in problems with the grasp when moving heavy items, only partially mitigated by the velocity scaling. If this would have been addressed explicitly earlier by the functional architecture, the motions could have been better designed to address this, e.g., maintain alignment of the nozzle with gravity to minimize detaching torque forces.

4.1 Concluding Remarks

Most teams participating in the Amazon challenge acknowledge that a development methodology focused on system-level integration and testing for early validation of design solutions that maximize their performance are key to develop such autonomous robots [5]. In this chapter, we have used the ISE&PPOOA for a system-level analysis of Team Delft's robot. This method allows to track the allocation of the different requirements in the system to the design decisions in the architecture of the solution. Besides, we propose a series of levels of robot automation to understand the fundamental properties of the different control patterns as design solutions to address uncertainty in the robot's operational scenarios. Furthermore, the use of MBSE approaches in the development of autonomous robots, when extended with formalisms to allow capturing these design decisions and its underlying rationale, result in formal models that capture the functional relation between the architecture of the system and its mission requirements, and which can be exploited by the systems at runtime, providing for new levels of self-awareness [3].

Acknowledgements This work has received funding from the European Union's Seventh Framework Programme under grant agreement no. 609206, and Horizon 2020 research and innovation programme under grant agreement no. 732287. The authors are very grateful to all their team members, and supporting sponsors and colleagues. Special thanks to G. van der Hoorn for his help during the development of the robotic system, and to Prof. J.L. Fernandez-Sanchez for the discussions on the design of robotic applications from a systems engineering perspective.

References

1. Eppner, C., Höfer, S., Jonschkowski, R., Martín-Martín, R., Sieverling, A., Wall, V., Brock, O.: Lessons from the Amazon Picking Challenge: four aspects of building robotic systems. In: Robotics: Science and Systems XII (2016)
2. Hernandez, C., Fernandez-Sanchez, J.L.: Model-based systems engineering to design collaborative robotics applications. In: 2017 IEEE International Systems Engineering Symposium (ISSE), pp. 1–6 (2017). https://doi.org/10.1109/SysEng.2017.8088258
3. Hernández, C., Bermejo-Alonso, J., Sanz, R.: A self-adaptation framework based on functional knowledge for augmented autonomy in robots. Integr. Comput. Aided Eng. **25**(2), 157–172 (2018)
4. Hernandez Corbato, C., Bharatheesha, M., van Egmond, J., Ju, J., Wisse, M.: Integrating different levels of automation: lessons from winning the Amazon robotics challenge 2016. IEEE Trans. Ind. Inf. **14**, 4916–4926 (2018). https://doi.org/10.1109/TII.2018.2800744
5. Morrison, D., Tow, A.W., McTaggart, M., Smith, R., Kelly-Boxall, N., Wade-McCue, S., Erskine, J., Grinover, R., Gurman, A., Hunn, T., Lee, D., Milan, A., Pham, T., Rallos, G., Razjigaev, A., Rowntree, T., Vijay, K., Zhuang, Z., Lehnert, C., Reid, I., Corke, P., Leitner, J.: Cartman: the low-cost cartesian manipulator that won the Amazon robotics challenge. In: 2018 IEEE International Conference on Robotics and Automation (ICRA), pp. 7757–7764 (2018). https://doi.org/10.1109/ICRA.2018.8463191
6. Parasuraman, R., Sheridan, T.B., Wickens, C.D.: A model for types and levels of human interaction with automation. IEEE Trans. Syst. Man Cybern. A Syst. Hum. **30**(3), 286–297 (2000). https://doi.org/10.1109/3468.844354
7. Prats, M., Şucan, I., Chitta, S., Ciocarlie, M., Pooley, A.: Moveit! workspace analysis tools (2013). http://moveit.ros.org/assets/pdfs/2013/icra2013tutorial/ICRATutorial-Workspace.pdf
8. Thrun, S., Montemerlo, M., Dahlkamp, H., Stavens, D., Aron, A., Diebel, J., Fong, P., Gale, J., Halpenny, M., Hoffmann, G., Lau, K., Oakley, C., Palatucci, M., Pratt, V., Stang, P., Strohband, S., Dupont, C., Jendrossek, L.E., Koelen, C., Markey, C., Rummel, C., van Niekerk, J., Jensen, E., Alessandrini, P., Bradski, G., Davies, B., Ettinger, S., Kaehler, A., Nefian, A., Mahoney, P.: Winning the DARPA grand challenge. J. Field Robot. (2006).

Standing on Giant's Shoulders: Newcomer's Experience from the Amazon Robotics Challenge 2017

Gustavo Alfonso Garcia Ricardez, Lotfi El Hafi, and Felix von Drigalski

1 Introduction

Incentivizing the development of innovative solutions to real-world problems through competitions has become a popular means for many companies to gather new, diverse ideas [18]. The Amazon Robotics Challenge (ARC), formerly known as the Amazon Picking Challenge, focuses on warehouse automation and has become one of the most renowned robotics challenges. The challenge aims to combine the state-of-the-art robotic manipulation and computer vision technologies into practical solutions for warehouse operation.

With increasingly stricter rules, it models the problem of stocking and picking items, which may not be known in advance. For reference of scale, Amazon introduces 50,000 new items every day to warehouses and fulfills an estimated 35 orders per second. At the ARC 2016, the winning team achieved a success rate of nearly 84%, approaching the performance of human workers, who pick and stow about 400 items per hour at full speed with an almost 100% success rate [1].

In this chapter, we present our approach to the ARC 2017, with particular emphasis on the lessons learned from past participants, our design philosophy and our development strategy. Moreover, we describe two elements of our proposed system, the suction tool and the shelf or *storage system*, and their development process.

The remainder of this chapter is structured as follows: First, we summarize the ARC 2017 tasks and technical challenges in Sect. 2. In Sect. 3, we present our

G. A. Garcia Ricardez (✉) · L. El Hafi · F. von Drigalski
Team NAIST-Panasonic at the Amazon Robotics Challenge 2017, Nagoya, Japan

Graduate School of Information Science, Nara Institute of Science and Technology (NAIST), Ikoma, Nara, Japan
e-mail: garcia-g@is.naist.jp; lotfi.el_hafi.kx2@is.naist.jp; felix.von_drigalski.fp6@is.naist.jp

A. Causo et al. (eds.), *Advances on Robotic Item Picking*,
https://doi.org/10.1007/978-3-030-35679-8_8

approach, the lessons learned from past competitors, our design philosophy, and our development strategy. In Sect. 4, we show an overview of our system and then present the details of the suction tool and storage system development. Finally, Sect. 5 concludes this chapter while highlighting the best lessons we learned.

2 Technical Challenges

The ARC consists of two tasks:

- *Stow*: Store 20 new items in the storage system, modeling the process of adding newly arrived items into the warehouse.
- *Pick*: Move 10 items from the storage system to three boxes, modeling the purchase process of Amazon.

The rules of the ARC 2017 have significant differences compared to previous years [2]:

(a) Half of the items are unknown until 30 min before the round starts.
(b) The storage system is designed by the teams.
(c) The volume of the storage system is 70% smaller than the previous years, for a total of 95 L.

(a) limits the applicability of conventional learning-based approaches in which a classifier is trained with large amounts of data (e.g., up to 150,000 images/item) to recognize items, which affects the approaches of eight of the top ten teams of the ARC 2016 [20, 21]. This requirement is realistic for warehouse applications, where new items are scanned and entered into the database and must be manipulated shortly after. For reference, Amazon enters about 50,000 new items to its inventory every day.

With (b), Amazon opened up a new design dimension in the challenge, allowing the teams to adapt the storage solution to their robot and to propose new ideas for the storage system. As the number of items remained the same as in previous years, the reduction of volume in (c) almost inevitably causes items to be stacked and occlusions to occur, which poses a significant challenge for object recognition, manipulation, and planning.

In summary, the main challenges are:

- **Object recognition:** Half of the items are unknown until 30 min before each round starts, which constrains following approaches where large amount of data can be used.
- **Robot manipulation:** The target items belong to very diverse categories (book, box, cylinder, deformable, wrapped, clamp-shell, and others), have

(continued)

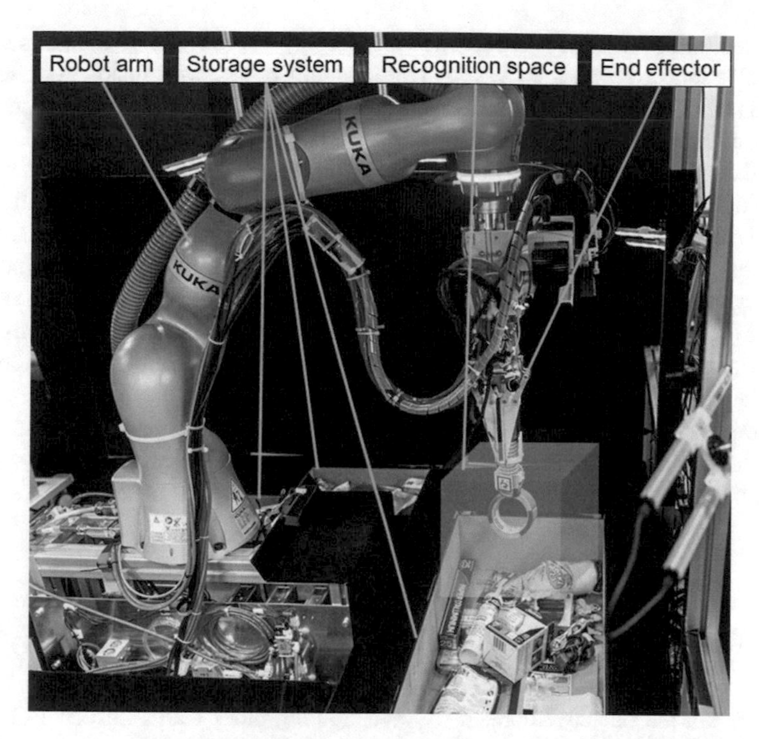

Fig. 1 Proposed warehouse automation robotic solution deployed by Team NAIST-Panasonic at the ARC 2017. The system consists of a 7-DOF robot arm, a custom-made end effector, a weight-aware storage system, and a recognition space. Each picked item is examined in the recognition space equipped with four RGB-D cameras to quickly recognize items from multiple viewpoints using both the state-of-the-art learning-based and feature-based technologies. Photo courtesy of Amazon Robotics

a maximum weight of 2 kg, and fit in a maximum volume of $0.42 \times 0.27 \times 0.14\,\text{m}^3$.

- **Storage system design:** Each team designs a storage system to fit a maximum of 32 items within 95 L of volume and $5000\,\text{cm}^2$ of area. It has to feature 2–10 bins, contain no actuators, a maximum of 50 USD in sensors (if any) and should be used for both tasks.

3 Approach

In general, our strategy consists of getting an initial estimation of the target item using the end effector camera, then pick it and move it to the *recognition space*, a dedicated and controlled space as shown in Fig. 1, where multiple methods vote to determine its class. Our idea was to create a democratic sensor fusion system using RGB-D cameras [5, 16], as well as weight, contact, and force sensors to identify all items and reduce uncertainty to a minimum [11].

In the stow task, we move items from a container to the storage system relying on the deep learning classification and weight, and discard erroneous classifications with the item list provided before the round and the already moved items. In the pick task, we (1) search for known items which are recognizable by the deep learning classifier, and (2) explore the bins by grasping unknown items and moving them to the recognition space to be identified. Finally, we move the items to its corresponding box, if it is a target item, or to another bin or location inside the same bin otherwise.

In the rest of this section, we present the logic behind our development and design decisions during the ARC 2017. We also present a survey on the experiences of past competitors, describe our design philosophy, and present the development strategy that we followed.

3.1 Past Competitions

We have started our development by investigating other teams' efforts in the two previous editions of the competition, which are instructive both in terms of how some approaches succeeded and what went wrong for others. In the remainder of this section, we detail past work that helped us gain insights into past editions of the ARC, and how this knowledge shaped our own design approach.[1]

A number of reports and media coverage have summarized the state of the art of ARC, as well as the accumulated heuristics, such as [10] by Team RBO who took first place in ARC 2015.

Correll et al. [6] describe platforms, grippers, sensors, and perception and motion planning techniques used by the teams competing in the 2015 edition. They conclude that there is trade-off between customization and dependability of software developed by the teams and third parties.

This was complemented by an in-depth report from Nikkei [20, 21] about the solutions of the 2016 teams. Additional reports from the Robotics Society of Japan illuminated some more approaches and problems in [9, 22].

A number of previous competitors, such as Team C^2M [19], Team R U Pracsys [28], and Team MIT-Princeton [32], also provide implementations of their approaches, as well as the datasets generated during the competitions. These datasets have been useful as a starting point to train and test our object recognition algorithms, as many of the items are also found in the ARC 2017 practice kit.

[1] While this section summarizes our literature review preparing for the ARC 2017, we note that papers detailing the ARC 2017 results have been presented at the 2018 ICRA workshop on *Advances in Robotic Warehouse Automation* [3], and/or awarded *Amazon Robotics Best Paper* [14].

Table 1 Common failures and the potential impact on the performance

Failure	Potential impact
Collision with storage system	Round loss
Planning failure	Round loss
Items left on recognition space	Object recognition capability loss
Losing suction contact	Point loss due to dropped item
Two-item grasping	Point loss due to lost item
Object recognition errors	Point loss due to misplaced item
Grasping failures	Time loss
Slow path-planning	Time loss

Looking at the past competitions, it becomes clear that using suction cups and deep learning has tended to increase teams' success rates. Furthermore, the reports show that teams using a single robot manipulator perform better, and make a strong case for reusability by using the Robot Operating System (ROS) [24, 25].

We have identified the most common problems that have occurred during the competition and summarized their potential impact in Table 1.

With these failures in mind, we drew the following main conclusions to guide our development effort:

- Suction is an effective grasping tool, as 80% of the items are suctionable.
- A professional suction system is important for reliable operation.
- Learning-based object recognition can yield up to 90% success rate [20, 21].
- Using depth information does not improve object recognition significantly, and may even be counterproductive.
- Robust error recovery is fundamental for a competitive performance.
- A 7-DOF manipulator can save time by achieving the target pose quicker as demonstrated by the previous winners [10, 13].
- Task planning using state machines is effective [4].
- Modifying the code in the last minute must be avoided, as it leads to human error.
- Sensors can overheat and stability issues should be anticipated.
- Illumination in the venue significantly affects the object recognition performance.

3.2 Design Philosophy

Our design philosophy is centered around simplicity and reliability. As there is no miracle solution for vision and manipulation in unstructured environments, we focus on designing dependable systems that tolerate errors, recover from failures, and continue the tasks safely.

The concept of separation of functions allows the specialization of components and supports clean design [31]. In our solution, we separated the functions of the end effector into suction and gripper tools, as shown in Fig. 2. The suction and gripper tools are mounted on separate linear actuators, which are used to advance and retract each tool before the manipulation. This arrangement allows separating their functions and increases both versatility and redundancy, as the implementation of per-tool as well as per-item manipulation strategies become simpler.

From the mechanical design to the software implementation, we harden our system against errors and implement recovery strategies. We implement error avoidance and recovery strategies on both the hardware and software level to make the system more robust. For example, if the drawer gets stuck and cannot be closed, the linear actuator shakes the drawer to make the blocking items fall inside and avoid massive penalties in the score. Though less likely but critical errors are hard to recover from, protocols for quick restart should always be prepared.

We use learning-based methods such as YOLO v2 [26, 27] and feature-based methods such as color histogram, bounding box volume, weight, and histogram of

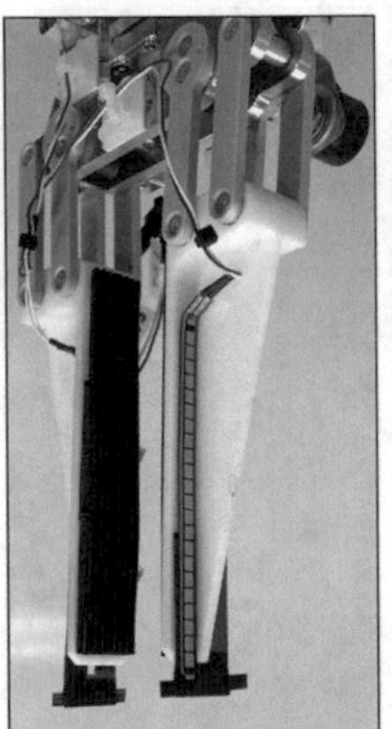

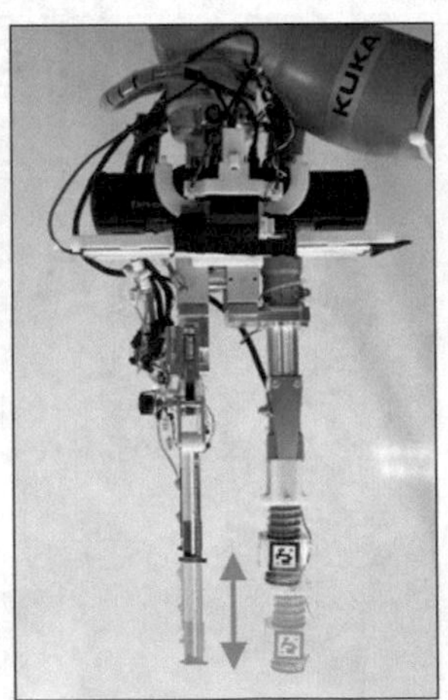

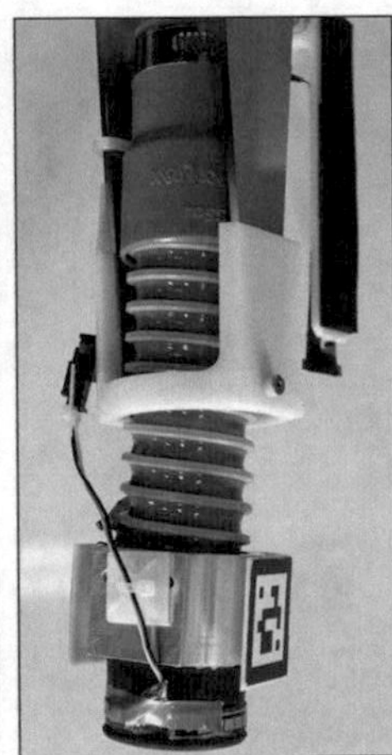

Fig. 2 The end effector (middle) consists of a suction tool (right) and gripper tool (left), each of which can move. The protruding hose contracts upon suction, which pulls the aluminum suction cup casing against the yellow ABS ring. Two markers on the casing allow the suction cup to be located easily during the recognition phase. A force-sensitive resistor (FSR) on the outside of the suction cup acts as a contact sensor. On the gripper, two protruding FSRs detect contact, and sensors underneath the rubber surface measure the contact force

oriented gradients [7] to detect and classify the items. The usage of multiple sensing technologies and methodologies helped to compensate for the weaknesses of each. For example, combining weight and volume by training a support vector machine yielded high accuracy in the recognition.

3.3 Development Strategy

In a first step, we aim to develop a minimal system that performs under limited and controlled conditions, so that we can test the basic functionality. We then iterate through short design and prototyping cycles, testing new ideas as early in the development process as possible.

In terms of hardware, we use the KUKA LBR iiwa [17] with a 14 kg payload, which has torque sensors in each of its seven joints. We use 3D printed parts and aluminum frames that are quickly accessible, versatile, and durable, and Arduino microcontrollers for both prototyping and deployment.

In our solution, the robot uses motion primitives to move between known configurations, as well as a planner [29] to pick up and deliver the items. This improves the overall performance of the robot, simplifies the high-level planning, and avoids getting stuck if planning solutions are time-consuming.

We use ROS to simplify the development, facilitate the integration, and increase the code reusability. To control the robot arm, we use the *iiwa_stack* [12] package to interface ROS and the KUKA API. We also use Git [30] version control and Docker [8] containers, which make it easy to share code and development environments across multiple devices and programmers. This helped us to quickly recover from an accidental deletion of source code and data that might have had a significant impact in our score, if no such tools have been employed.

The source code of our solution is publicly available online in the official ARC repository.[2]

4 Proposed Solution

The proposed system consists of a custom-made end effector mounted on a serial robotic manipulator, a controlled recognition space formed by an array of RGB-D cameras, and a storage system with weight sensors. The end effector features suction and gripper tools for manipulation, as well as an RGB-D sensor for object recognition and grasping point estimation. An overview of the components of the system is shown in Fig. 3.

[2]Warehouse Picking Automation Challenges, Team NAIST-Panasonic: https://github.com/warehouse-picking-automation-challenges/team_naist_panasonic.

In this section, we focus on describing the development of the storage system and the suction tool, which was the most used manipulation tool since 80% of the items are graspable by suction.

4.1 Suction Tool

The suction tool consists of a compliant, partially-constrained vacuum cleaner hose of 150 mm in length and 35 mm in diameter connected to an aluminum tube of 150 mm length and 35 mm in diameter. The tip of the suction tool is made of flexible rubber and has a force-sensitive resistor to detect contact with the items. The suction tool dimensions were determined by analyzing the suction force required to lift the maximum possible weight of an item set in the ARC rules as well as the flow over force ratio to ensure a reliable suction seal.

As shown in Fig. 2, several centimeters of the compliant hose extend beyond the fixation. This compliant connection allows the suction cup to move and incline, thus compensating for angular and position errors as the suction pulls the cup and object together. Furthermore, it enables the robot to move quickly into the containers, without danger of damaging articles during a collision. Lastly, when the suction seal is successful, the hose contracts, which pulls the suction cup casing and the suctioned item towards the fixation, making the formerly compliant connection more rigid and reducing the item movement (e.g., swinging) after suction.

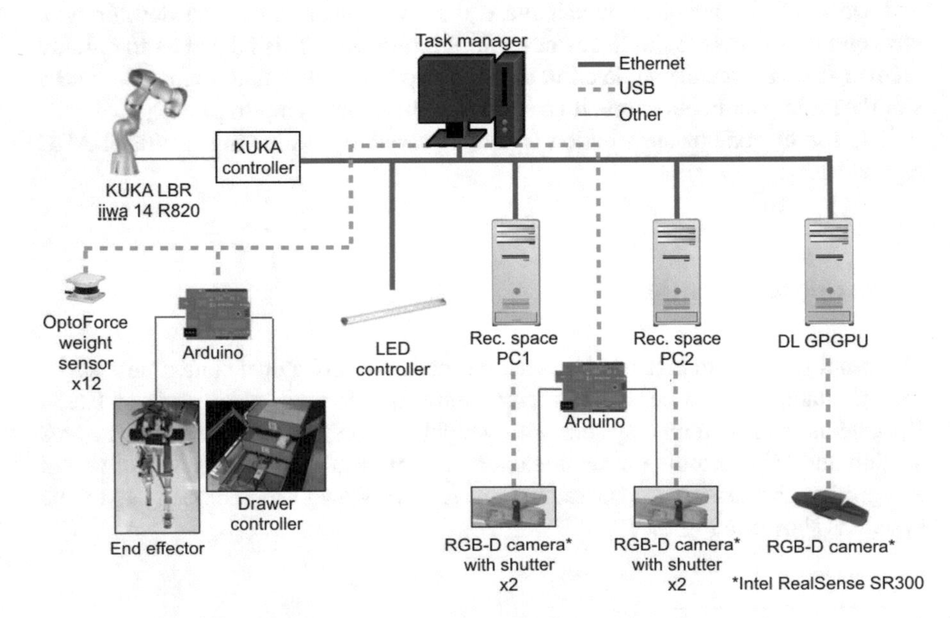

Fig. 3 Overview of the Team NAIST-Panasonic solution deployed at ARC 2017

In past competitions, teams have struggled with designing a suction mechanism with sufficient suction force. Commercial vacuum cleaners and similar solutions do not generate the necessary flow and pressure difference to secure all items. On the other hand, excessive suction force may damage the packaging of the item, e.g., clothes in PVC bags. To respond to this problem and to design a system that can suction all items safely, we have investigated the suction force systematically.

First, we modeled the suction tool as shown in Fig. 4 as a long tube with an opening at the end. Then, with the pressure difference and flow rate of the blower, the hose diameter, the suction cup size, and the relative opening at the end (assuming an imperfect seal), we calculate the resulting normal force. We performed preliminary experiments with suction cups of 30, 40, and 50 mm in diameter d_p, and hoses of 10, 20, 30, 40, and 50 mm in diameter d and 5 m in length L. Figure 5 shows the results.

The combination of $d_p = 40$ mm and $d = 30$ mm had the best performance when tried with all items from the ARC 2017 practice kit: 36 items can be suctioned (90%). The suction force has to be controlled because nine of the items (22.5%) can be damaged by excessive suction force. Thin and cylindrical items (e.g., wine glass, toilet brush, dumbbell) and the mesh cup were the most challenging to pick via suction. Nonetheless, it is notable that even porous, deformable, and irregularly shaped items such as the marbles and the body scrubber could be picked reliably.

The main conclusion is that with sufficient hose diameter and air flow, items can be held even if the suction seal is imperfect, such as when the item surface is uneven or rough. With smaller hose sizes, suction cups of smaller diameters break away significantly earlier, as they cannot transport enough air to sustain a leaking seal.

We decided to use an industrial grade blower [15] with a maximum vacuum of −40 kPa to power the suction tool. We use a pressure sensor to detect when an item has been suctioned and if the item has been dropped (i.e., when the suction seal has broken). To avoid damaging the delicate packaging of items, we added a waste gate with a PD controller to the hose that can regulate the static pressure at the suction cup between 0 and −40 kPa.

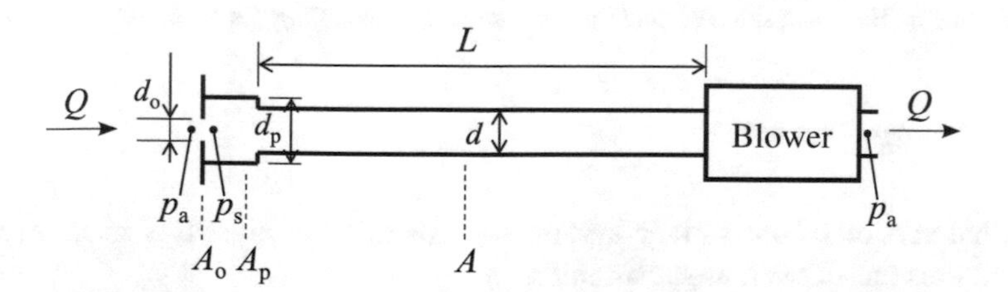

Fig. 4 Simplified model of the suction tool to systematically investigate the suction force. Q is the flow rate, p_a and p_s are the external and internal static pressures, respectively. A, A_p, and A_o are the cross sections of the tube, the suction cup chamber and the suction cup border, respectively, and their corresponding diameters are denoted with d. L is the length of the hose

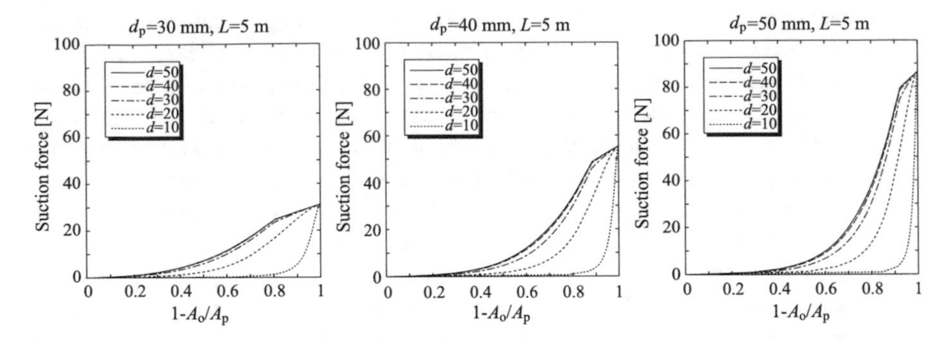

Fig. 5 Effective suction force for an imperfect suction seal, assuming the vacuum machine Induvac VC 355-720. $1 - A_o/A_p$ is 1 for a perfect suction seal, and 0 when no part of the object is in contact. It can be seen that larger hose diameters make the suction seal more robust, as they allow higher air flow

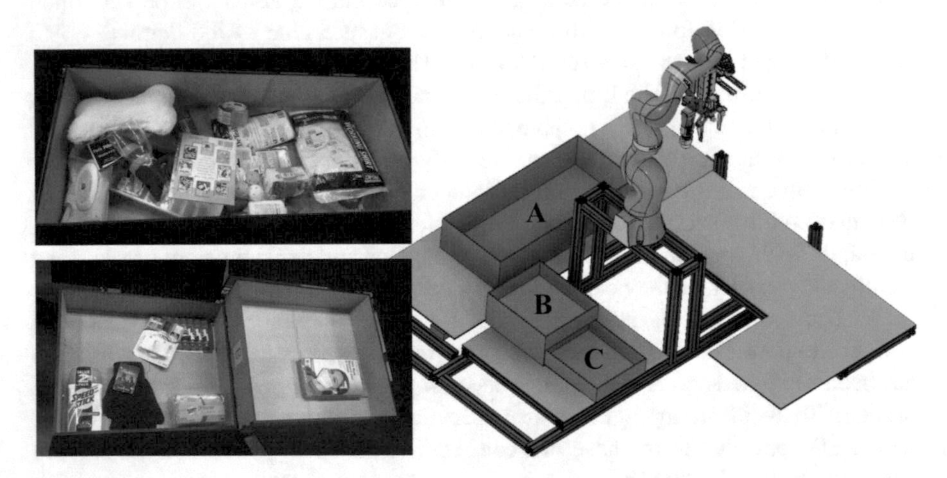

Fig. 6 Storage system featuring three bins. Bin A (top left) is designed to store large items. Bin B and C (bottom left) are shallower to store small items. Bin C is a drawer opened and closed by a linear actuator. The black skirts around the bins prevent items falling out of the storage area

4.2 Storage System

The final version of our storage system consists of three bins. It is made of wood and features one drawer, as shown in Fig. 6.

We started our development with a storage system composed of adjustable aluminum profiles. In retrospect, the ability to adjust division sizes was unnecessary, as we did not end up testing many different configurations. For the first prototype, we fixed the depth of the shelf to accommodate the second largest dimension of the item maximum size, i.e., 27 cm. The other two dimensions were set to

Table 2 Dimensions of the bins used in the horizontally oriented storage system

Name	Length (m)	Width (m)	Depth (m)
Bin A	0.90	0.40	0.19
Bin B	0.40	0.30	0.08
Bin C	0.38	0.29	0.08

an approximated square, making the shelf compliant with the rule of 95,000 cm^3 maximum volume.

During the preliminary tests with this initial design, we realized that it becomes cluttered very easily, and requires precise manipulation to avoid items falling out. We experimented with different configurations and drawers to decrease the clutter and finally opted for a horizontally oriented storage system with three bins, as described in Table 2. It is worth mentioning that most successful teams included a system that increased the available surface area of the storage system to spread out items and reduce clutter.

We aimed to maximize the surface available to place items considering both the maximum allowed volume and the kinematic reachability of the manipulator. To this end, we designed bin C as a drawer which is opened and closed by a linear actuator attached to the robot structure. This drawer holds small items which otherwise would be difficult to find in an unstructured pile of items.

We placed two sets of weight sensors: one under bin A and another under bins B and C. Each set of weight sensors consists of four 3D force sensors [23] attached to a rectangular base made of aluminum frames and can detect changes of up to 2 g. Taking into account the considerable effort and cost to implement a custom weight sensing solution, as well as the hysteresis effects in the force sensors that negatively affect overall precision, we would strongly recommend choosing an off-the-shelf alternative.

The data from the weight sensors is used to discriminate between events such as (un)successful grasping and accidental drops, as well as to classify the items by weight or a combination of weight and other features. Actually, we combined the weight of the item measured by the force sensors, and the bounding box volume calculated in the recognition space with a Support Vector Machine (SVM) for classification. Combining weight and volume not only allows the SVM to include a measure of density into its classification, which is especially helpful to differentiate the numerous light items, but also catches certain edge cases such as clamshell items (e.g., open books) effectively. On the other hand, the bounding box of highly deformable objects can be inconsistent, which makes them vulnerable to misclassification.

Finally, in order to avoid dropping items outside of the bins and to increase the chances of recovery, we installed ramps around the bins that help catch dropped items.

5 Conclusion

We presented our approach to the ARC 2017, with particular emphasis in the lessons learned from past participants, our design philosophy and our development strategy. We also described two particular features of our proposed system, namely the suction tool and the storage system, which we believe played an important role in our performance.

The main lessons from our experience in the ARC 2017 can be summarized as follows:

- Consider previous competitors' experience to avoid making similar mistakes.
- Use code and datasets from past competitors to get started.
- Keep it simple and only do what is necessary. Overengineering and unnecessary redundancy make the system more prone to fail.
- Use tools such as Git, Docker, and ROS to facilitate development and recovery from unexpected errors.
- Start with a minimal system that can be iterated upon quickly, to keep prototyping and development cycles short.
- Develop a robust error recovery as perception and manipulation errors as well as uncertainty are unavoidable.
- Avoid modifying the code in the last minute, as this leads to human error.
- Take into account illumination as this significantly affects the object recognition performance in the venue.
- Protect sensors against overheat and anticipate stability issues.
- Work on logistics from the beginning as transporting robots is often time-consuming.

Appendix

Table 3 shows the final results obtained by Team NAIST-Panasonic during the ARC 2017.

Table 3 Final results of Team NAIST-Panasonic at ARC 2017

Tasks	Results	Rank	Participants
Stow task	110 points	Fourth place	16 teams
Pick task	80 points	Tenth place	16 teams
Combined final	90 points	Sixth place	8 teams
Overall challenge	Finalist prize	Sixth place	16 teams

Acknowledgements This work derives from the participation of the Team NAIST-Panasonic at the Amazon Robotics Challenge 2017, Nagoya, Japan. The authors thank the team members for their valuable contributions during the competition: Gustavo Alfonso Garcia Ricardez (Captain, NAIST), Lotfi El Hafi (NAIST), Felix von Drigalski (NAIST), Pedro Miguel Uriguen Eljuri (NAIST), Wataru Yamazaki (NAIST), Viktor Gerard Hoerig (NAIST), Arnaud Delmotte (NAIST), Akishige Yuguchi (NAIST), Marcus Gall (NAIST), Chika Shiogama (NAIST), Kenta Toyoshima (NAIST), Rodrigo Elizalde Zapata (NAIST), Seigo Okada (Panasonic), Pin-Chu Yang (Panasonic), Yasunao Okazaki (Panasonic), Katsuhiko Asai (Panasonic), Kazuo Inoue (Panasonic), Ryutaro Futakuchi (Panasonic), Yusuke Kato (Panasonic), and Masaki Yamamoto (Panasonic). The authors also thank Thomas Denison for his technical assistance, and Salvo Virga for his work on iiwa_stack and tireless technical support. Finally, we would like to thank Amazon for hosting the ARC, which has continued to advance the state of the art of both research and open source software.

References

1. Ackerman, E.: Team Delft wins Amazon Picking Challenge. IEEE Spectr. (2016). https://spectrum.ieee.org/automaton/robotics/industrial-robots/team-delft-wins-amazon-picking-challenge
2. Amazon: Amazon Robotics Challenge (2016). https://www.amazonrobotics.com/#/roboticschallenge/rules
3. Amazon: Amazon Robotics best paper announcement (2018). https://www.amazonrobotics.com/site/binaries/content/assets/amazonrobotics/pdfs/ar-best-paper-announcement.pdf
4. Bohren, J., Cousins, S.: The SMACH high-level executive. IEEE Robot. Autom. Mag. **17**(4), 18–20 (2010)
5. Carfagni, M., Furferi, R., Governi, L., Servi, M., Uccheddu, F., Volpe, Y.: On the performance of the Intel SR300 depth camera: metrological and critical characterization. IEEE Sensors J. **17**(14), 4508–4519 (2017)
6. Correll, N., Bekris, K.E., Berenson, D., Brock, O., Causo, A., Hauser, K., Okada, K., Rodriguez, A., Romano, J.M., Wurman, P.R.: Analysis and observations from the first Amazon Picking Challenge. IEEE Trans. Autom. Sci. Eng. (T-ASE) **15**(1), 172–188 (2018)
7. Dalal, N., Triggs, B.: Histograms of oriented gradients for human detection. In: 2005 IEEE Conference on Computer Vision and Pattern Recognition (CVPR 2015), vol. 1, pp. 886–893 (2005)
8. Docker: Docker engine (2013). https://www.docker.com/
9. Domae, Y.: Technology trends and future about picking and manipulation by robots. J. Robot. Soc. Jpn. **35**(1), 13–16 (2017)
10. Eppner, C., Hofer, S., Jonschkowski, R., Martin-Martin, R., Sieverling, A., Wall, V., Brock, O.: Lessons from the Amazon Picking Challenge: four aspects of building robotic systems. In: 26th International Joint Conference on Artificial Intelligence (IJCAI 2017), Melbourne, pp. 4831–4835 (2017)
11. Garcia Ricardez, G.A., von Drigalski, F., El Hafi, L., Okada, S., Yang, P.C., Yamazaki, W., Hoerig, V.G., Delmotte, A., Yuguchi, A., Gall, M., Shiogama, C., Toyoshima, K., Uriguen Eljuri, P.M., Elizalde Zapata, R., Ding, M., Takamatsu, J., Ogasawara, T.: Warehouse picking automation system with learning- and feature-based object recognition and grasping point estimation. In: 2017 SICE System Integration Division Annual Conference (SI 2017), Sendai, pp. 2249–2253 (2017)
12. Hennersperger, C., Fuerst, B., Virga, S., Zettinig, O., Frisch, B., Neff, T., Navab, N.: Towards MRI-based autonomous robotic US acquisitions: a first feasibility study. IEEE Trans. Med. Imaging (T-MI) **36**(2), 538–548 (2017)

13. Hernandez, C., Bharatheesha, M., Ko, W., Gaiser, H., Tan, J., van Deurzen, K., de Vries, M., Van Mil, B., van Egmond, J., Burger, R., Morariu, M., Ju, J., Gerrmann, X., Ensing, R., Van Frankenhuyzen, J., Wisse, M.: Team delft's robot winner of the Amazon Picking Challenge 2016. arXiv Robotics (cs.RO), pp. 1–13 (2016)
14. IEEE: Advances in robotic warehouse automation (2018). http://juxi.net/workshop/warehouse-automation-icra-2018/
15. Induvac: Double stage side channel blowers 3 phase VC 355-720 (2010). http://www.induvac.com/en/downloads/data-sheets/side-channel-blowers/
16. Intel: RealSense Camera SR300 (2016). https://software.intel.com/en-us/realsense/sr300
17. KUKA: LBR iiwa 14 R820 (2013). https://www.kuka.com/en-de/products/robot-systems/industrial-robots/lbr-iiwa
18. MacCormack, A., Murray, F., Wagner, E.: Spurring innovation through competitions. MIT Sloan Manag. Rev. **55**(1), 25–32 (2013)
19. MPRG: Team C^2M: APC RGB-D+PointCloud Dataset 2015 (2015). http://mprg.jp/research/apc_dataset_2015_e
20. Nikkei: Amazon Picking Challenge 2016 in-depth report (Part 1). Nikkei Robotics **14**, 4–16 (2016)
21. Nikkei: Amazon Picking Challenge 2016 in-depth report (Part 2). Nikkei Robotics **15**, 16–22 (2016)
22. Okada, K.: Perspective from international robotics competitions. J. Robot. Soc. Jpn. **35**(1), 9–12 (2017)
23. OnRobot: OptoForce OMD-20-FG-100N (2016). https://onrobot.com/products/omd-force-sensor/
24. Open Source Robotics Foundation: Robot operating system (2009). http://www.ros.org/
25. Quigley, M., Gerkey, B., Conley, K., Faust, J., Foote, T., Leibs, J., Berger, E., Wheeler, R., Ng, A.: ROS: an open-source robot operating system. In: 2009 IEEE Workshop on Open Source Software, Kobe (2009)
26. Redmon, J., Farhadi, A.: YOLO9000: better, faster, stronger. arXiv Computer Vision and Pattern Recognition (cs.CV), pp. 1–9 (2016)
27. Redmon, J., Divvala, S., Girshick, R., Farhadi, A.: You only look once: unified, real-time object detection. In: 2016 IEEE Conference on Computer Vision and Pattern Recognition (CVPR 2016), Las Vegas, pp. 779–788 (2016)
28. Rennie, C., Shome, R., Bekris, K.E., De Souza, A.F.: A dataset for improved RGBD-based object detection and pose estimation for warehouse pick-and-place. IEEE Robot. Autom. Lett. (RA-L) **1**(2), 1179–1185 (2016)
29. Sucan, I.A., Chitta, S.: MoveIt! (2011). https://moveit.ros.org/
30. Torvalds, L.: Git (2005). https://git-scm.com/
31. von Drigalski, F., El Hafi, L., Uriguen Eljuri, P.M., Garcia Ricardez, G.A., Takamatsu, J., Ogasawara, T.: Vibration-reducing end effector for automation of drilling tasks in aircraft manufacturing. IEEE Robot. Autom. Lett. (RA-L) **2**(4), 2316–2321 (2017)
32. Zeng, A., Yu, K.T., Song, S., Suo, D., Walker, E., Rodriguez, A., Xiao, J.: Multi-view self-supervised deep learning for 6D pose estimation in the Amazon Picking Challenge. In: 2017 IEEE International Conference on Robotics and Automation (ICRA 2017), Singapore, pp. 1386–1383 (2017)

Team C^2M: Two Cooperative Robots for Picking and Stowing in Amazon Picking Challenge 2016

Hironobu Fujiyoshi, Takayoshi Yamashita, Shuichi Akizuki, Manabu Hashimoto, Yukiyasu Domae, Ryosuke Kawanishi, Masahiro Fujita, Ryo Kojima, and Koji Shiratsuchi

1 Introduction

At the Amazon.com logistics warehouse in the USA, the Kiva Pod robot made by Kiva Systems (Amazon Robotics as of 2016) automatically conveys products from the storage shelves to the people responsible for picking. Manual labor is still needed to pick these products from the shelves, but it is expected that this task will eventually be automated by introducing picking robots. In an e-commerce business where there are many types of products stored randomly on shelves, the key to the introduction of automation is being able to perform stable pick-and-place operations by recognizing diverse objects on the shelves and gripping them correctly. Against this background, Amazon set up the Amazon Picking Challenge (APC) as a competitive event for robots in the automation of logistics. The first such event was APC 2015, which focused on the problem of picking diverse items. Contestants were required to build picking robots that could extract 25 different

H. Fujiyoshi (✉) · T. Yamashita
Chubu University, Kasugai, Aichi, Japan
e-mail: fujiyoshi@isc.chubu.ac.jp; takayoshi@isc.chubu.ac.jp

S. Akizuki · M. Hashimoto
Chukyo University, Nagoya, Aichi, Japan
e-mail: s-akizuki@sist.chukyo-u.ac.jp; mana@isl.sist.chukyo-u.ac.jp

Y. Domae
AIST, Koto-ku, Tokyo, Japan
e-mail: domae.yukiyasu@aist.go.jp

R. Kawanishi · M. Fujita · R. Kojima · K. Shiratsuchi
Mitsubishi Electric, Amagasaki, Hyogo, Japan
e-mail: Kawanishi.Ryosuke@bx.MitsubishiElectric.co.jp;
Fujita.Masahiro@aj.MitsubishiElectric.co.jp; Kojima.Ryo@bc.MitsubishiElectric.co.jp;
Shiratsuchi.Koji@cj.MitsubishiElectric.co.jp

A. Causo et al. (eds.), *Advances on Robotic Item Picking*,
https://doi.org/10.1007/978-3-030-35679-8_9

items from 12 frames (called "bins") on a shelf [1]. At APC 2016 in the following year, the scope of the competition was made more realistic by having the contestants compete on two different tasks ("Pick" and "Stow"). This paper introduces the robot system of Team C^2M at Amazon Picking Challenge 2016, and its image recognition system by grasping position based object recognition.

2 Two Cooperative Industrial Robots

In this section, we discuss the Team C^2M robot system and its features.

2.1 Robot System

As shown in Fig. 1, the Team C^2M robot system consists of two MELFA industrial robots with load capacities of 7 kg (RV-7FL) and 4 kg (RV-4FL), both equipped with 3D vision sensors (MELFA-3D Vision), force sensors (4F-FS001), and multifunctional hands. The robot with a 7-kg load capacity was installed on a single-axis sliding platform. The robot that directly accesses the shelf needs to be able to pick up randomly placed items. We used the robot with a 7-kg load capacity for this purpose because it has a longer reach and can carry heavier loads. For the robot that extracts items from the tote, we used a robot with a 4-kg load capacity, which has sufficient reach and load capacity for this purpose.

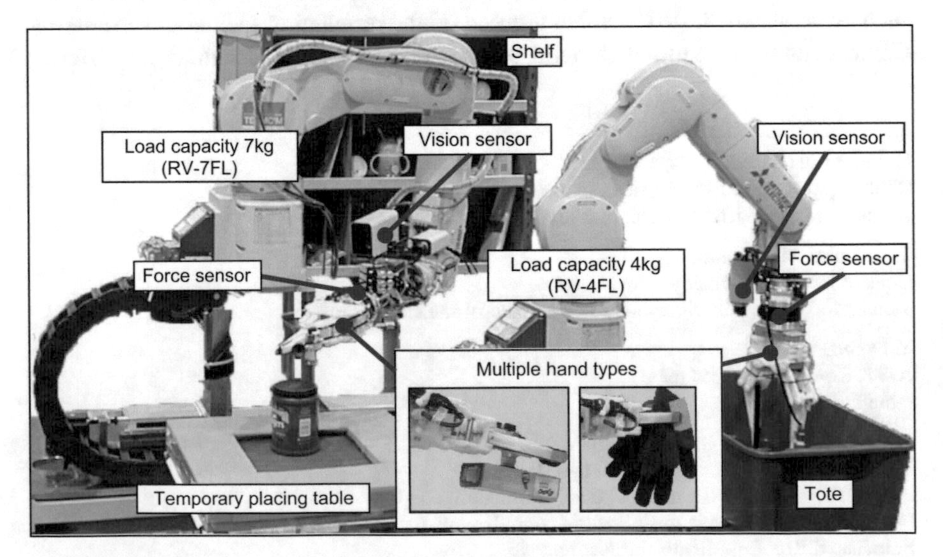

Fig. 1 Overall view of the Team C^2M robot system

A force sensor is mounted at the end of the robot arm to prevent collisions and breakages by judging if unnecessary force is being applied to the robot, shelf, or item when gripping products. Also, by touching the shelf with a hand equipped with a force sensor, we were able to judge the fine positional offset from the reference position, and implemented automatic calibration between the shelf and robot. The 3D vision sensors produced RGB image and depth image outputs by employing an active stereo method comprising a camera and projector. This sensor is used for three-dimensional measurement and recognition of items.

2.2 Features

When putting the items picked up by the robot on the shelf, it is necessary to estimate the item's 6D object pose and set it down with the optimal orientation. However, there are a wide variety of items to be picked including rigid objects, non-rigid objects, and transparent objects, and it is difficult to estimate the 6D object orientation of all these types of object. Therefore, our team stored items on the shelf by having two robots work cooperatively via a temporary placing table. By placing items temporarily on the temporary placing table, the robots can hand over the items in such a way that they are easy to put away. By introducing a temporary placing table between the two robots, we simplified the complex task of putting away items on the shelf.

3 Vision Strategy

Here, we describe vision strategy of the Team C^{2}M.

3.1 Grasping Position Based Object Recognition by CNN

A Convolutional Neural Network (CNN) identifies items that appear in the input images. In general, the recognition of objects by a CNN is performed as shown in Fig. 2a, where candidate object regions called "region proposals" are extracted and provided as input to the CNN for identification. To detect as many region proposals as possible to avoid missing out any picking items, this type of CNN must run many times, resulting in a large computational load.

To reduce this load, we used the approach shown in Fig. 2b, where the image is first analyzed to detect grasping positions, and the CNN is only then used to identify these items based on the image regions surrounding these detected grasping positions. By detecting the grasping positions first, we can limit the recognition processing to these detected positions. This approach is much more efficient because

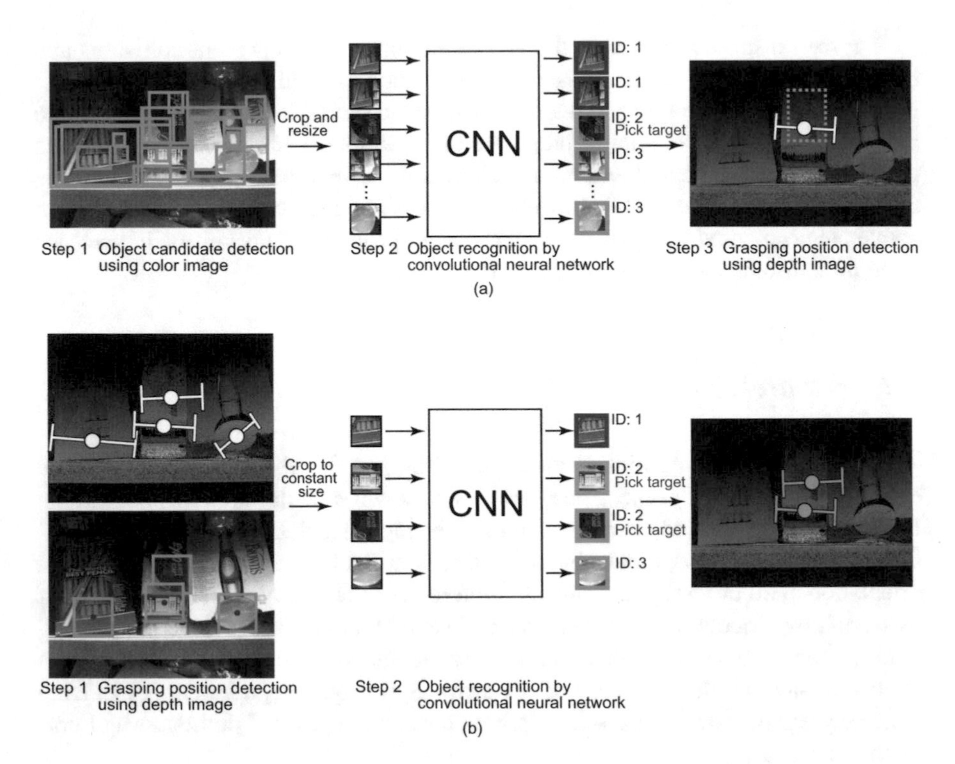

Fig. 2 Workflow of object recognition for pick-and-place unit. (**a**) Conventional object recognition for pick-and-place unit. (**b**) Proposed object recognition for pick-and-place unit

the robot can then move directly on to picking actions. The grasping positions can be detected at high speed by using Fast Graspability Evaluation [2].

As shown in Fig. 3, the CNN in the proposed method consists of convolution layers and fully connected layers. The convolution layers use batch normalization in layers $\{1, 2\}$ and max pooling in layers $\{1, 2, 5\}$. We used the ReLU activation function for all layers. In class identification by a CNN, the class probabilities are generally calculated by the *softmax* function. Here, the class probabilities are calculated by the softmax function $P_r(\cdot)$ as shown in Eq. (1), where h_i represents the values of the CNN output units and C is the number of classes.

$$P_r(h_i) = \frac{\exp(h_i)}{\sum_{j=1}^{C} \exp(h_j)} \tag{1}$$

In the task set by the Amazon Picking Challenge, the robot system received a JSON file containing the names of items stored in the bin, and the names of the items to be picked. The robot system thus knows what items are in the bin before it performs object recognition. We propose a *constrained softmax* function where the

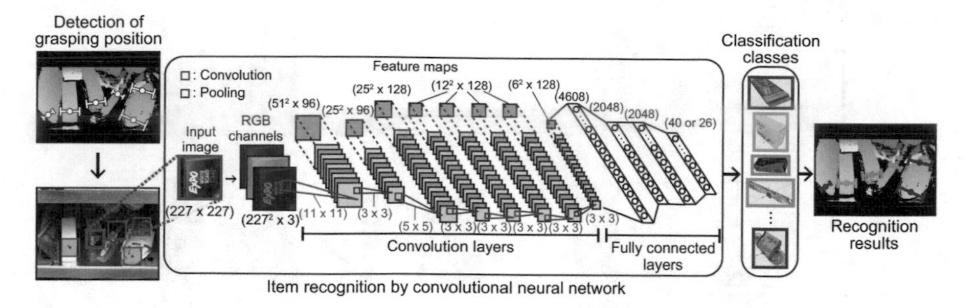

Fig. 3 CNN network structure

CNN output layer only calculates class probabilities for output units corresponding to items that are present in the bin. For example, when the bin contains items corresponding to the set of output units $\{1, 3, 4\}$, the constrained softmax function $P_c(\cdot)$ can be defined as shown in Eq. (2).

$$P_c(h_i) = \frac{\exp(h_i)}{\sum\limits_{j=\{1,3,4\}} \exp(h_j)} \tag{2}$$

Using the constrained softmax function, it is possible to eliminate the possibility of mistakenly recognizing items that are not in the bin.

3.2 *Picking Strategy*

The Pick task involves moving each of the twelve items in a bin by picking them up one at a time and putting them in a box called a "tote." In this task, a JSON file containing the names of the items in the bin is read in and applied to the constrained softmax function. Figure 4 shows the item recognition procedure. First, the target bin is photographed to acquire an image. Information about the items inside the bin is then read in from the JSON file. Next, the items to be picked are recognized by the constrained softmax function using only the output units that correspond to the items in the bin.

3.3 *Stowing Strategy*

The Stow task involves picking twelve randomly placed items from the tote and storing them in bins on the shelf. Figure 5 shows the item recognition procedure in the Stow task. In the Stow task, we used two robot arms and force sensors either side

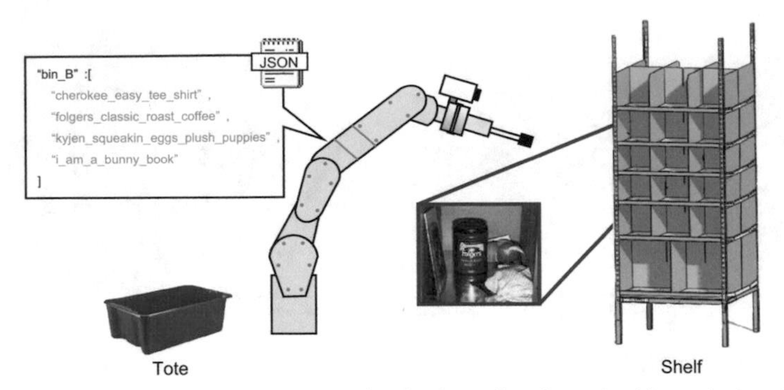

Fig. 4 Picking strategy

of a temporary placing table. First, we pick a graspable item from the tote. The item is moved to the temporary placing table, and its weight is measured by the force sensor. Based on the item's weight, the list of candidate graspable items is narrowed down to four possibilities. The item placed on the temporary placing table is then recognized by the constrained softmax function using output units corresponding to these four possibilities based on the item's weight.

4 Experiments

In this experiment, we compared the CNN's recognition precision based on the grasping position and its processing time of the R-CNN [3], the Faster R-CNN [4], and the proposed method. For "Region proposal" detection in R-CSS, "Selective search" [5] is used.

4.1 Dataset

The dataset used in this experiment are the products used by the APC. A total of 39 types of objects were used in APC 2016. Only images with a single object placed on the shelf are used for CNN learning. In APC 2016, 1709 learning images were used. For evaluation, we use 200 images of which multiple objects were placed on the shelf.

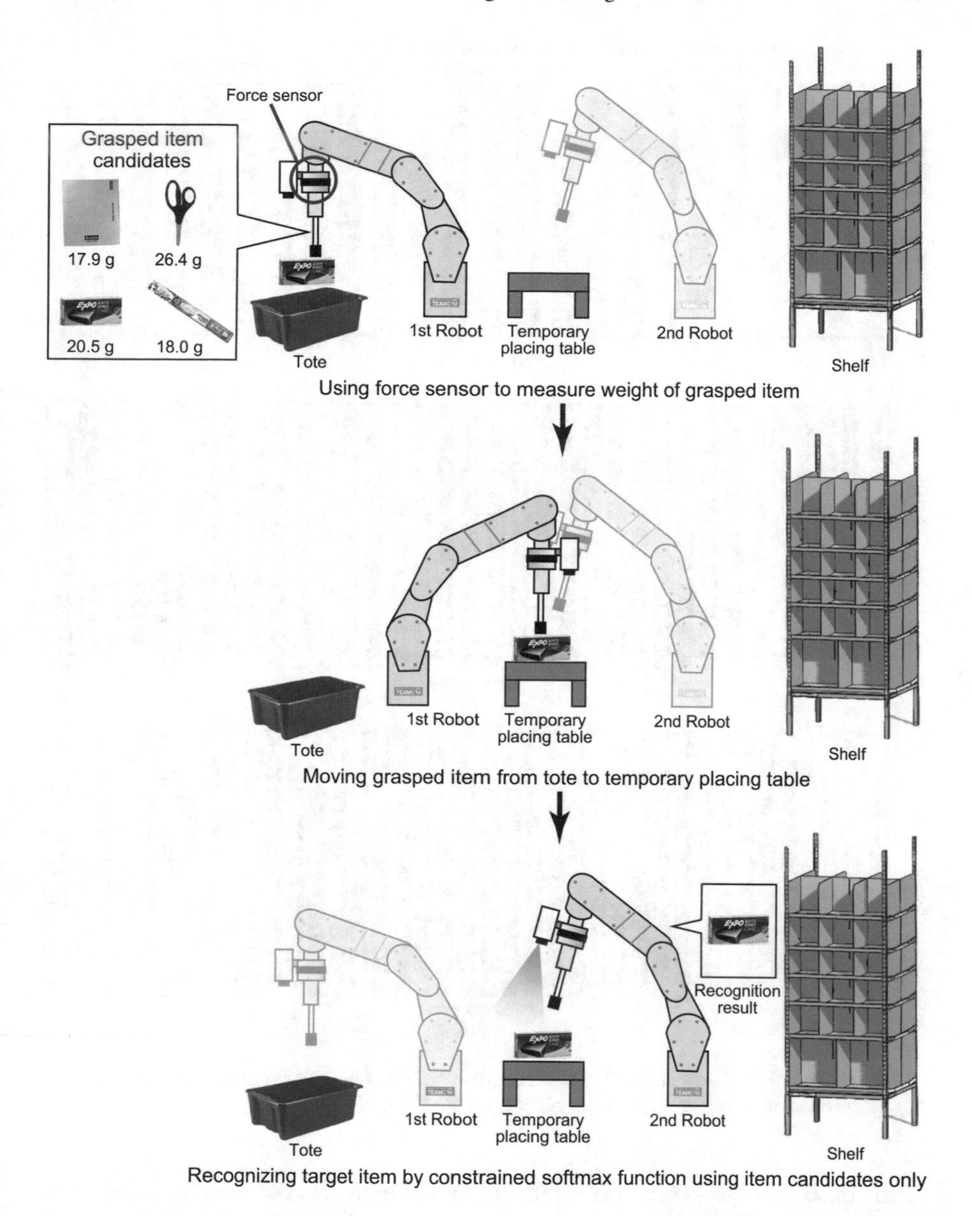

Fig. 5 Stowing strategy

4.2 *Accuracy for Object Recognition*

We compare CNN's recognition precision based on the grasping position in the R-CNN, Faster R-CNN, and the proposed method. We evaluated the APC 2016 dataset for three cases: 2–3 objects were placed on the shelf, 4–5 objects were placed on the

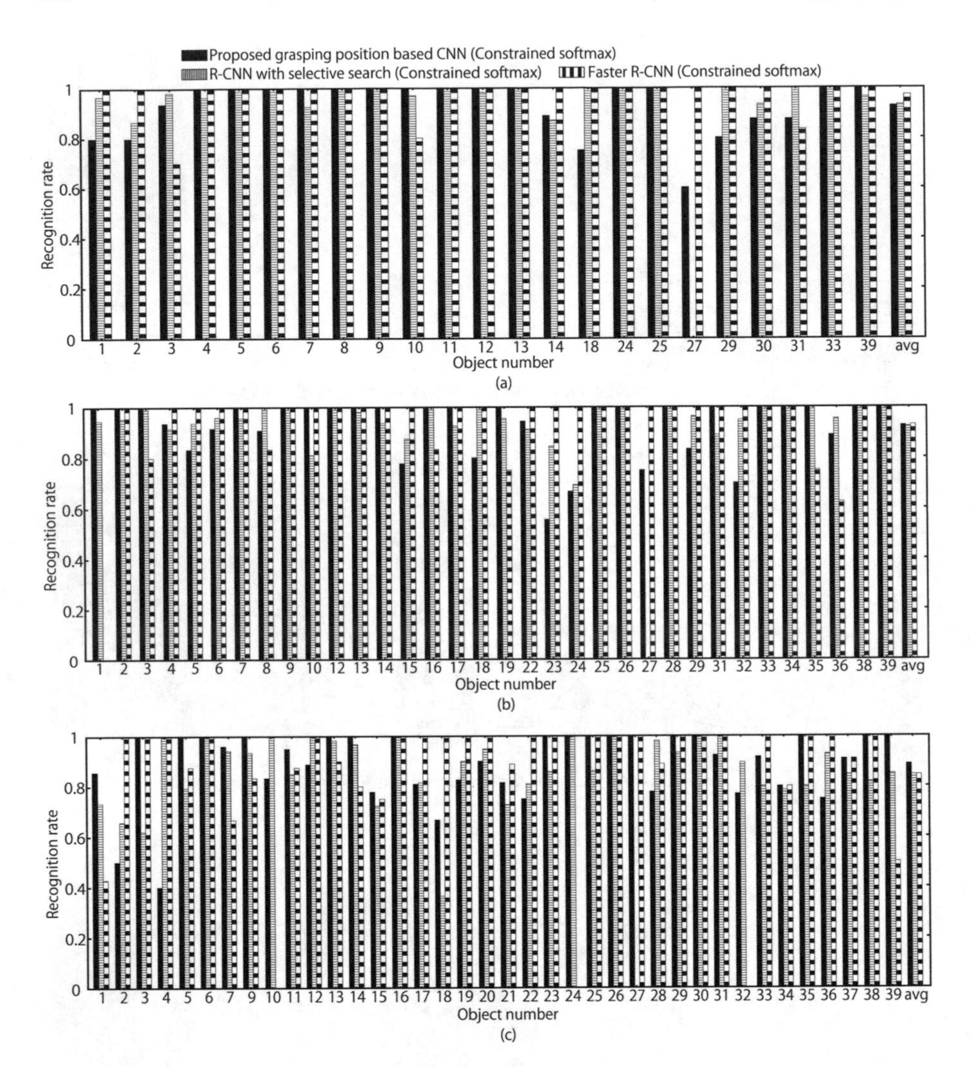

Fig. 6 Recognition rate of APC 2016 dataset. (**a**) Number of objects on shelf = 2–3. (**b**) Number of objects on shelf = 4–5. (**c**) Number of objects on shelf = 6–10

shelf, and 6–10 objects were placed on the shelf. Figure 6 shows the recognition rate of each object. The proposed method shows the average recognition rate, the same as the R-CNN's rate or better and, in the case where the number of objects was 6–10, it showed an average recognition rate 4.3% higher than the Faster R-CNN's rate.

Lastly, Fig. 7 shows the object recognition result of our team (Team C^2M) in the Amazon Picking Challenge 2016 held in Leipzig, Germany in July 2016. The competition consisted of two tasks—"Pick task" and "Stow task." The pick task required picking a specified object from a shelf and transferring it to a box called a "tote." The stow task required storing an object on the shelf. Figure 7 shows the

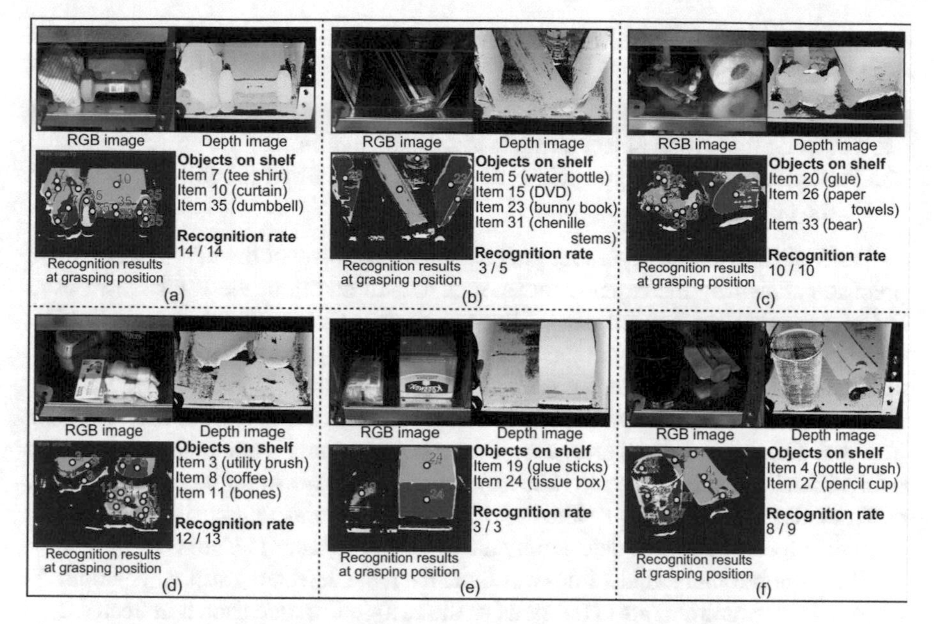

Fig. 7 Recognition results for our team in Amazon Picking Challenge 2016. (**a**) Scene 1 (bin I), (**b**) Scene 2 (bin B), (**c**) Scene 3 (bin H), (**d**) Scene 4 (bin E), (**e**) Scene 5 (bin J), (**f**) Scene 6 (bin L)

recognition results of six scenes where the proposed method (pick task) tried to recognize the objects. In the pick task, we can directly apply the constrained softmax function because the JSON file can be used where the names of the objects stored in the shelf are described. The recognition rate in the pick task is 92.5% – 50/54 (= Correct recognition result/Detected grasping position), achieving high-precision object recognition for an actual problem. The stow task transfers the object picked up from the "tote" to a temporary platform, and the proposed method identified the object on the temporary platform. The stow task measures the weight of the object grasped in the "tote" by the force sensor installed in the robot, without using the JSON file. By selecting four types of candidates from the weight of the object, you can use the softmax function to recognize the object on the temporary platform; the stow task achieved high-precision recognition.

4.3 Accuracy for Grasping Position Detection

This section compares the accuracy rates of grasping position detection after object recognition. Section 4.2 compared the recognition rates of R-CNN and Faster R-CNN for object rectangle detected. The methods (R-CNN, Faster R-CNN, etc.) based on "Region proposal" need to detect the grasping position in the rectangular region detected after detecting the position of the recognized object as a rectangular region. The R-CNN or Faster R-CNN may grasp the wrong object for the following two reasons.

1. Grasping the wrong object because of false recognition of the object rectangle.
2. False detection of the grasping position because a different object is projected in the object rectangle.

The first cause is that the robot grasps the wrong object at the rate of recognition precision shown by the experiment in Sect. 4.2. In addition, the R-CNN or Faster R-CNN invokes the second cause. Even though the class of the detected object rectangle is correct, if many objects are placed close together on the shelf as shown in Fig. 8, the robot may detect a different grasping position in the object rectangle.

On the other hand, the proposed method recognizes the object for each grasping position. The reason why it grasps the wrong object is only when it has made a false recognition. Table 1 shows the accuracy rate for each grasping position candidate when the Fast Graspability Evaluation [2] detects the grasping position candidates from the object rectangles detected by the R-CNN or Faster R-CNN.

The proposed method identifies an object from the detected grasping position. As a result, its recognition rate is the same as the recognition rate shown in Sect. 4.2. In addition, in all methods the constrained softmax function is applied for only objects to be detected on the shelf. If there are a large number of objects on the shelf, the R-CNN or Faster R-CNN may have to detect multiple objects in one rectangle, significantly reducing the accuracy rate of the grasping position. The proposed

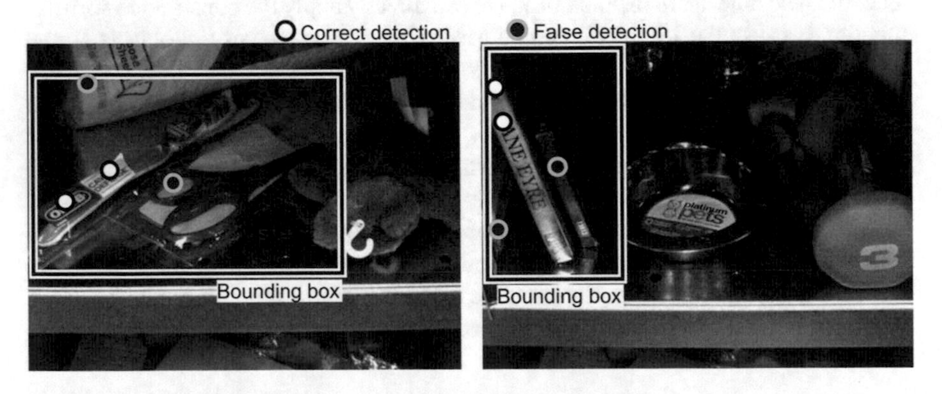

Fig. 8 Examples of grasping position within detected bounding box by Faster R-CNN

Table 1 Accuracy rate of grasping position within bounding box [%]

# of objects	Proposed grasping position based CNN	R-CNN with selective search	Faster R-CNN
2–3	92.72	79.97	91.51
4–5	92.69	56.85	82.49
6–10	88.89	55.61	74.54

Table 2 Breakdown of object recognition time [ms]

	Proposed grasping position based CNN		R-CNN with selective search		Faster R-CNN	
	CPU	GPU	CPU	GPU	CPU	GPU
Selective search	–		1662.3		–	
Det. of grasping positions	579.6		570.1		559.4	
Inference processing of CNN	245.1	8.9	622.8	26.8	10,551.9	69.2
Total	824.7	588.5	2855.2	2259.2	11,111.3	628.6

method achieves the highest precision in all datasets where there are three or more objects on the shelf.

4.4 Computational Time

We compared the average processing time for those 30 images. We measured the processing time by using "CPU: Intel Core i7-7700 3.6–4.2 GHz" and "GPU: NVIDIA GeForce GTX 1080 Ti." The number of average grasping positions was five in the proposed method. The number of average object candidates was 14. In all methods, the Fast Graspability Evaluation was used to detect the grasping position. Table 2 shows a breakdown of the processing time needed for recognition.

Table 2 shows that the processing time of the CPU in the proposed method is faster by about 3.4 times that of R-CNN and is faster by about 13.4 times that of Faster R-CNN. The R-CNN has a very large ratio of processing time to detect "Region proposal" by "Selective search." Furthermore, "Selective search" excessively detects more object candidates because it reacts to the texture (i.e., characters, logos, etc.) printed on the objects, increasing the number of CNN executions and delaying the processing time. The Faster R-CNN processes three tasks (i.e., "Region proposal detection", "Object rectangular regression", "Object classification") in one CNN and requires a large-scale network compared to the CNN in the proposed method. As a result, a large processing time is required to execute the CNN. Even using a GPU, the processing time for the proposed method is the shortest. The proposed method can recognize multi-class objects quickly and efficiently.

5 Conclusion

Although APC 2016 featured harder problems than APC 2015, the problem scenarios were still somewhat artificial due to characteristics such as the low incidence of occlusion between items. Since real problems are more complex, it is expected that future events will have a higher level of difficulty in both the Pick task and the Stow task in order to approximate real-world problems more closely. We also felt the need to concentrate more on robot safety, following an accident where someone's arm was injured by colliding with a robot hand during the competition. Picking and storing items more flexibly is likely to be a key subject for future study. The purpose of the Amazon Picking Challenge is to analyze and discuss technical issues by building robot systems, analyzing their performance, and competing them against one another. Although there are still various issues that must be overcome before such systems become a practical reality, the open sharing of knowledge at this event aims to promote the resolution of key issues. Through the Amazon Picking Challenge, it is hoped that large advances will be made in picking robot technology.

References

1. Correll, N., Bekris, K.E., Berenson, D., Brock, O., Causo, A., Hauser, K., Okada, K., Rodriguez, A., Romano, J.M., Wurman, P.R.: Lessons from the Amazon Picking Challenge. Preprint. arXiv: 1601.05484 (2016)
2. Domae, Y., Okuda, H., Taguchi, Y., Sumi, K., Hirai, T.: Fast graspability evaluation on single depth maps for bin picking with general grippers. In: International Conference on Robotics and Automation, pp. 1997–2004 (2014)
3. Girshick, R., Donahue, J., Darrell, T., Malik, J.: Rich feature hierarchies for accurate object detection and semantic segmentation. In: Conference on Computer Vision and Pattern Recognition, pp. 580–587 (2014)
4. Ren, S., He, K., Girshick, R., Sun, J.: Faster R-CNN: towards real-time object detection with region proposal networks. In: Advances in Neural Information Processing Systems, pp. 91–99 (2015)
5. Uijlings, J.R., Van De Sande, K.E., Gevers, T., Smeulders, A.W.: Selective search for object recognition. Int. J. Comput. Vis. **104**(2), 154–171 (2013)

Description of IITK-TCS System for ARC 2017

Anima Majumder, Olyvia Kundu, Samrat Dutta, Swagat Kumar, and Laxmidhar Behera

1 Introduction

The Amazon Robotics Challenge (ARC) is being organized by Amazon to spur research and innovation in the area of warehouse robotics and automation. The competition started in 2015 focusing only on the pick task where the robot had to pick items from a rack with multiple bins and place them to a tote. The stow task was included in 2016 where, in addition to picking, the robot had to move items from a tote to a given rack. The problem was further made difficult in 2017 by introducing several complexities in pick and stow tasks. For instance, the rack dimensions were reduced substantially thereby increasing the clutter in the rack that one would encounter during pick or stow operation. Similarly, the participants were provided with 50% of new products during the actual competition making it impossible to use pre-trained models. In addition, the diversity of products used for the competition was further increased by including difficult objects such transparent glass goblets, soft and deformable items, clothes, a heavier metallic dumb-bell, large folders, large books with soft cover, etc. Given this wide diversity, it is not possible to pick all the items using either suction or gripping alone and hence provides ample scope of innovation in gripper design itself.

In this paper, we provide the details of our system that was used for participation in the ARC 2017 competition. Some of the novel contributions made by our team are as follows. The first contribution is an automated annotation system that could

A. Majumder (✉) · O. Kundu · S. Dutta · S. Kumar
TATA Consultancy Services, Bangalore, India
e-mail: anima.majumder@tcs.com; olyvia.kundu@tcs.com; samrat.dutta2@tcs.com; swagat.kumar@tcs.com

L. Behera
Indian Institute of Technology, Kanpur, India
e-mail: lbehera@iitk.ac.in

A. Causo et al. (eds.), *Advances on Robotic Item Picking*,
https://doi.org/10.1007/978-3-030-35679-8_10

generate a large number of labeled templates in a short interval of time and train a deep network to recognize and segment objects in a clutter. The details of the system is described in Sect. 2. The second contribution lies in designing and developing a new hybrid gripper that combines a suction action with a two-finger gripping action. The details of this gripper and other hardware components are discussed in Sect. 3. The third contribution is a new geometry based grasping algorithm that can compute graspable affordances of objects in a cluttered environment. The details of the algorithm is discussed in Sect. 4. We provide some discussion on robot calibration, motion planning, and some performance analysis in Sects. 5, 6, and 8, respectively, followed by conclusion in Sect. 9.

2 Object Recognition Algorithms

In the ARC competition, robot was required to recognize objects in a clutter before performing pick or stow task. Most of the teams used deep learning to solve the object recognition problem [1, 2]. Since these methods required large amount of data and time to train deep networks, each team was provided with the object set well advance in time (several months) before the competition. The problem was further made difficult in 2017 by providing 50% new items just 45 min prior to the competition. So, each team had to generate templates and train their deep network within this stipulated time. We solved this problem by creating a hardware rig as shown in Fig. 1 and using a deep learning framework for object recognition. The platform consists of a rotating platform and a set of cameras mounted at varying height and view angles. The object is placed on the rotating base and images are stored at regular intervals of the rotation. This allowed us to generate a large number of templates (about 300 images per object) in a very short interval of time. The image processing method required for generating labeled templates from these images is shown in Fig. 2. It uses a semi-supervised deep convolutional network for both annotation as well as final pixel-wise classification for 40 objects provided for the competition. We use a standard ResNet based deep network that is used as both a

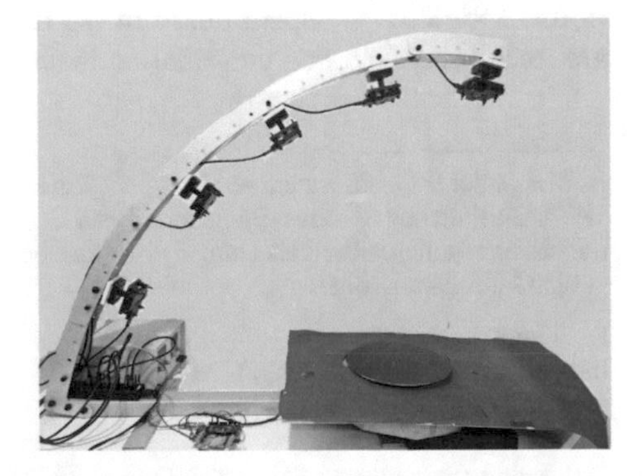

Fig. 1 The hardware rig used for automating the data generation process

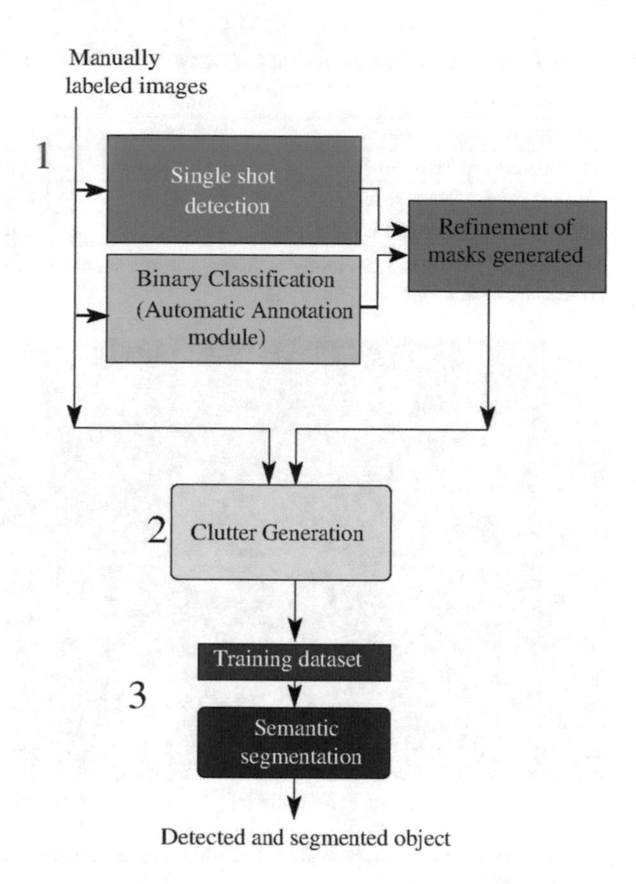

Fig. 2 Deep Learning based method for generating automatic annotations

binary classifier and a multi-class classifier as explained below. This network uses a Feature Pyramid Network (FPN) that aggregates multi-level features obtained from successive layers of the standard ResNet network to provide better segmentation between objects without needing additional data and time, making it faster than other deep network methods.

The proposed method comprises of three steps. First, the above deep network is trained on a small set of manually annotated dataset to act as a binary classifier that can segment a foreground object from its background. This binary classifier is fine-tuned using a rectangular ROI detector based on single shot detection (SSD) technique [3] to generate masks for each object and store them as labeled templates. The second step involves generating clutters synthetically by superimposing individual templates obtained from the previous step. Third step involves training the above deep network using this machine generated clutter dataset for multi-class segmentation. This proposed framework for automatic data generation as well as object segmentation is shown to be more accurate and faster compared to the existing state-of-the-art methods like Masked-RCNN [4] and PSPNet [5] as shown in Table 1. The better performance of our proposed framework can also be

Table 1 Performance comparison of deep networks with ResNet-50 as the base network

Performance measure	Mask R-CNN	PSPNet	Proposed net
Average pixel accuracy	90.34%	95.64%	96.76%
Average forward pass time	115.20 ms	108.066 ms	63.622 ms
Average backward pass time	192.8 ms	172.048 ms	82.650 ms
Average forward-backward pass time	308 ms	280.460 ms	146.598 ms

The comparison is made in terms of Average pixel accuracy of segmentation, average time for processing each image

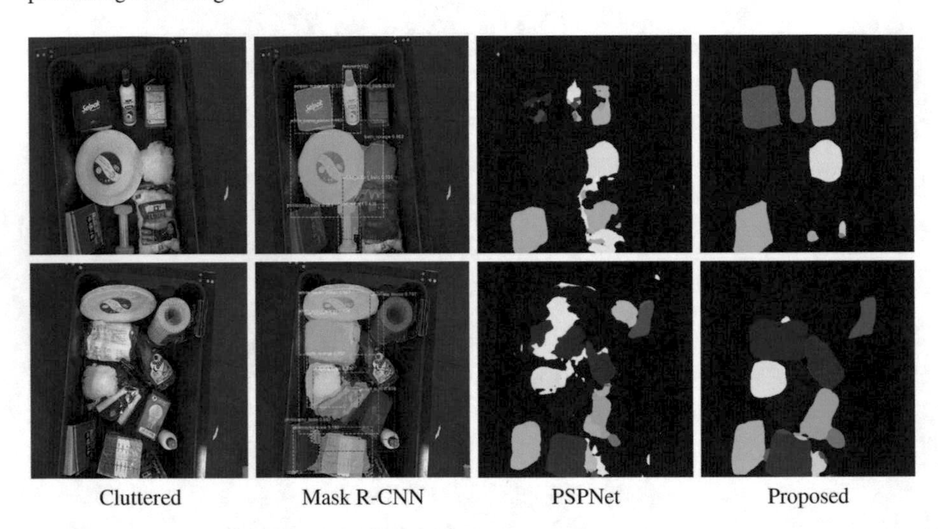

Fig. 3 Segmentation comparison for real cluttered images using Mask R-CNN, PSPNet, and the Proposed architecture

corroborated from Fig. 3 where it is shown to provide better segmentation compared to other two state-of-the-art methods. The complete details of the method is being omitted here in order to maintain the brevity of this chapter. Interested readers may refer to the published articles [6] and [7] for more details.

3 Custom Gripper Design

As described earlier, it is not possible to grasp all kinds of objects using only suction or gripping action alone. To address this problem, we designed a hybrid gripper mechanism that combines suction system with a two-finger gripper as shown in Fig. 4. The suction system has a swivelling bellow cup at its end that can rotate by 90° in the vertical plane. The two-finger gripper has a maximum clearance of about 11 cm between the fingers. It uses a retractable mechanism to slide forward and backward on the top of a long horizontal suction beam. This hybrid gripper uses

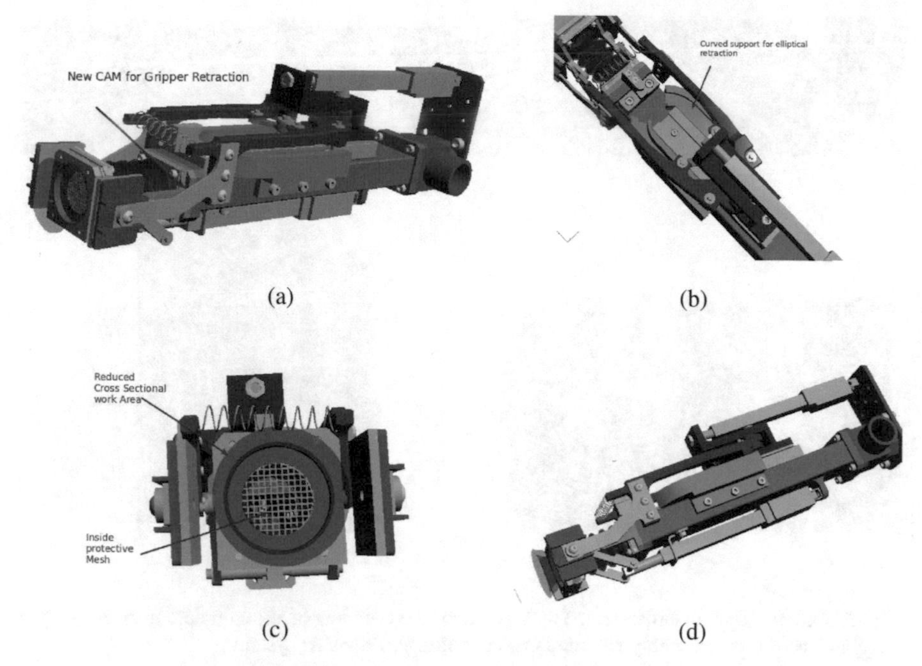

(a) (b) (c) (d)

Fig. 4 CAD diagrams for the hybrid gripper. It combines vacuum based suction system with a retractable two-finger gripper. Linear actuators with multi-link cam mechanisms are used for reducing the form factor of the gripper. (**a**) CAM for gripper retraction. (**b**) Elliptical support for gripper retraction. (**c**) Suction end of the gripper. (**d**) Two linear actuators

two linear actuators for actuating all kinds of motion. It is designed for a payload of 2 kg and weighs around 1.5 kg. The total length of gripper is about 40 cm and has a cross section of about 8 × 8 cm in its retracted position as shown in Fig. 4c. It has been designed to work in constrained workspaces. The actual hardware is shown in Fig. 5a, b. The complete gripper assembly on the robot arm along with camera and lighting mechanism is shown in Fig. 5c, d. A video demonstration showing the actual working of the gripper is being made available online [8] for the convenience of readers.

A vacuum system is used to generate suction for picking up objects. However, it creates a lag while releasing the object. The problem is solved by using a valve to generate a negative vacuum. The valve is shown in Fig. 6b, d which uses a linear actuator as a piston to release the pressure inside the pipe and thus, helps in reducing the drop time significantly. A flow sensor shown in Fig. 6a is used to detect failures while picking the object by measuring the pressure difference.

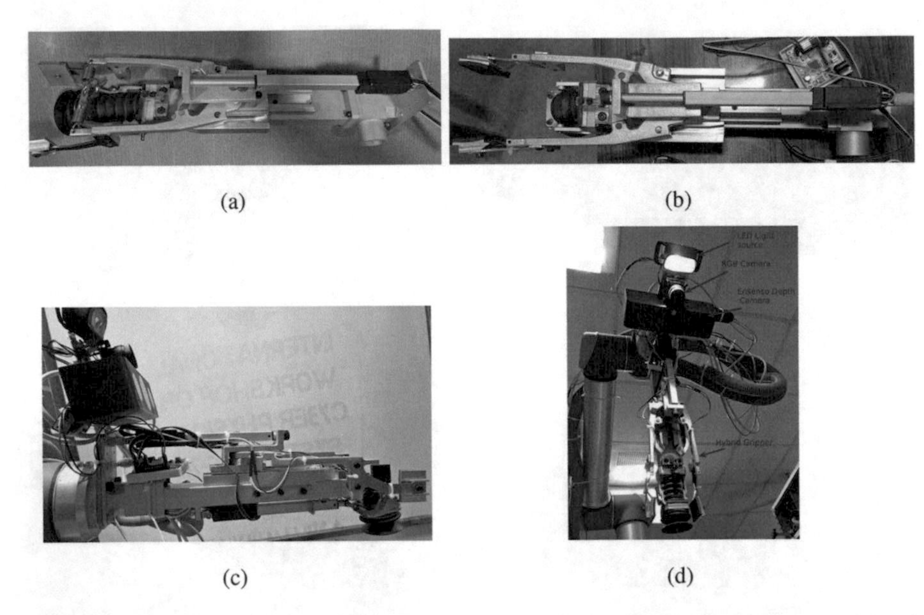

(a) (b)

(c) (d)

Fig. 5 Actual Gripper Hardware. (**a**) and (**b**) show the top view of the complete gripper. (**c**) and (**d**) show the gripper assembly mounted on robot along with vision system

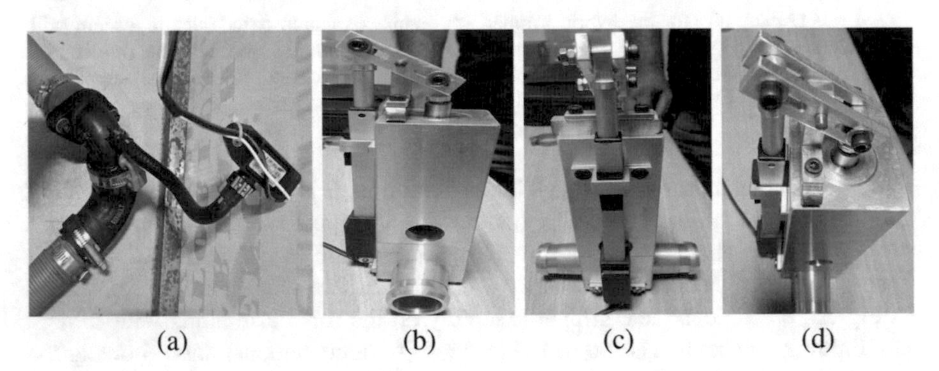

(a) (b) (c) (d)

Fig. 6 Pressure release mechanism to reduce drop time. (**a**) Flow sensor is used to detect pressure inside the suction tube. (**b**)–(**d**) Multiple views of the pressure release device that creates a negative vacuum to allow the objects drop faster once the suction is stopped

4 Grasping Algorithm

In order to successfully pick or place objects, it is important to detect suitable graspable affordances which can be used by robot grippers to hold these objects. It is also known as grasp pose detection [9] and is considered to be a difficult problem in the robotics research community. We take a geometric approach to detect graspable affordances directly in a 3D point cloud obtained from a range sensor. It primarily involves two steps. The first step uses surface continuity of surface normals to

Fig. 7 Finding graspable affordances in extreme clutter using a novel geometry based method

identify the natural boundaries of objects by using a modified version of the region-growing algorithm that uses a pair of thresholds and a concept of *edge points* to remove spurious edges and thereby, identifying natural boundaries of each object even in a clutter. Once the surfaces for different objects are segmented, the second step uses the gripper geometry to localize the graspable regions. The six dimensional grasp pose detection problem is simplified by making practical assumption of the gripper approaching the object in a direction opposite to the surface normal of the centroid of the segment with its gripper closing plane coplanar with the minor axis of the segment. The principal axes for each segment (major, minor and normal axes) are computed using Principal Component Analysis (PCA). Essentially, the valid grasping regions are localized by carrying out a one-dimensional search along the principal axes of the segment and imposing the geometrical constraints of the gripper. This is made possible by projecting the original 3D Cartesian points of the surface onto the principal axes of the surface segment. The outcome of this approach is shown in Fig. 7. As one can observe, we are able to detect multiple graspable affordances for each object (shown as multi-color bands along with corresponding pose axes). We are able to detect grasping handles for rectangular objects like books which are difficult compared to curved surfaces. Readers are referred to [10] for the details of the algorithm which is being omitted here in order to meet the page constraints for this chapter.

5 Robot-Camera Calibration

Robot-camera calibration is required mapping visual information seen through a camera to real world coordinates that can be reached by the robotic arm. The calibration process involves estimating various transformation matrices required for carrying out this mapping from image space to the real world robot workspace. We use a combination of RGB camera and an Ensenso N35 depth camera [11] in an *eye-in-hand* configuration as shown in Fig. 5c, d, respectively. In order to facilitate further discussion, let us define the four coordinate frames as follows: (1) ***RGB-frame***: the coordinate frame of the RGB camera, (2) ***c-frame***: the coordinate frame of the Ensenso depth camera, (3) ***e-frame***: the coordinate frame for the robot end-effector and, (4) ***b-frame***: coordinate frame at robot based, also known as the global frame. Any point $\mathbf{p}^{RGB}$ seen in the RGB camera frame (*RGB-frame*) can be transformed into a point $\mathbf{p}^b$ in the robot base frame using the following equation:

$$\mathbf{p}^b = T_e^b \cdot T_c^e \cdot T_{RGB}^c \cdot \mathbf{p}^{RGB} \tag{1}$$

where the homogeneous transformation matrix $T_b^e = \{T_e^b\}^{-1}$ transforms a point in the robot base frame (*b-frame*) to the end-effector frame (*e-frame*) and is obtained from the robot forward kinematic equations. The matrix T_c^e transformations a point in depth camera frame (*c-frame*) to robot end-effector frame (*e-frame*). This is estimated by solving a least square problem between a set of points seen in the camera frame and their corresponding points recorded from the robot by making the end-effector to touch the same points in the workspace. The transformation matrix T_{RGB}^c is required for associating each pixel coordinate in RGB image plane with a corresponding depth value obtained from the Ensenso depth camera. This matrix is estimated by using the camera calibration toolbox available in the OpenCV library [12].

6 Motion Planning with Obstacle Avoidance

We are primarily using *Moveit!*[1] package available with ROS for developing motion planning algorithms. The simulation is carried out using Gazebo[2] Simulator. The motion planning problem is solved in two main stages: (a) joint space motion planning and (b) Cartesian space motion planning. Joint level planning is required for fixed robotic postures such as home position of the robot, bin view pose. During a stow/pick task execution, the robot starts from the home position and goes to the tote/bin view pose first and looks for the targeted object. These are fixed joint positions which are planned in joint level. Joint space motion planning is quite simple as compared to the Cartesian space planning. An $S - curve$ trajectory is generated from the current joint position to the final joint position, so that the trajectory maintains a trapezoidal velocity profile. The length of the linear portion of the $S - curve$ depends on the speed of operation. In contrast, the Cartesian space motion planning comprises three steps: (1) selection of the target pose of the gripper, (2) Cartesian trajectory generation from the current pose of the gripper to the target pose, and (3) finding inverse kinematics for each points in the trajectory. This is discussed in the following section.

The target gripper pose for a given object is obtained from the perception module. The object recognition and localization module provides the centroid ($\mathbf{t}^c \in \Re^3$), the major axis ($\mathbf{a}_{major}^c \in \Re^3$), and the surface normal vector ($\mathbf{n}^c \in \Re^3$) for a given

[1]https://moveit.ros.org/.

[2]http://gazebosim.org/.

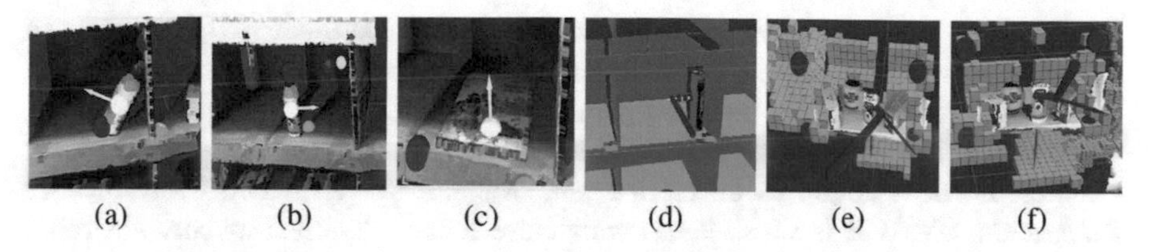

(a) (b) (c) (d) (e) (f)

Fig. 8 Grasp pose detection under different conditions (**a**–**d**). (**e**) and (**f**) show the path generated by the motion planner using OctoMap representation for obstacles

object visible in the camera frame. This triplet $\{\mathbf{t}^c, \mathbf{a}^c_{major}, \mathbf{n}^c\}$ specified in *c-frame* is transferred into the robot base frame (*b-frame*) by using the transformation matrices $\mathbf{T}^e_c$ and $\mathbf{T}^b_e$ and Eq. (1). This transformed triplet $\{\mathbf{t}^b, \mathbf{a}^b_{major}, \mathbf{n}^b\}$ is used to construct the rotation matrix needed for defining the desired pose of the robot end-effector. This rotation matrix is defined by

$$\mathbf{R}^b = \left[\mathbf{n}^b \ \mathbf{a}^b_{major} \ \mathbf{n}^b \times \mathbf{a}^b_{major}\right] \tag{2}$$

This rotation matrix along with the centroid of the object surface constitutes the target pose used by the Moveit motion planning library to generate trajectory in the Cartesian robot workspace. Some of the target poses for objects thus computed is shown in Fig. 8a–c. The Moveit motion planning algorithm uses flexible collision library (FCL) [13] to generate paths for robot arm that avoid collision with the rack as well as the objects surrounding the target item. In order to avoid collision with the rack, the bin corners are used to define primitive boxes for each wall of the bin. These primitive boxes are then treated as obstacles in the motion planning space. Similarly, the collision with other objects in the bin is achieved by creating 3D occupancy map called OctoMap [14] which converts point cloud into 3D voxels. This OctoMap feed is added to the Moveit motion planning scene and FCL is used for avoiding collision with the required objects. Once the pose is detected properly, the motion planning algorithm is used for finding the suitable path for reaching the target. Some of the path generated by Moveit motion planner is shown in Fig. 8d–f. Once the end-effector trajectory is obtained in the Cartesian workspace, the corresponding joint angle trajectories are generated by solving the robot inverse kinematics equation using the built-in TRACK-IK library [15] of Moveit package. These joint angle trajectories are then fed to the robot controller for execution.

7 Software Architecture for Implementation

Most of the algorithms for various components were developed using C/C++, Python on Linux operating system. The system integration was carried out using ROS framework. It uses standard ROS topics, services, and parameters to exchange data and command over multiple machines connected to each other by an Ethernet-based LAN. The execution of tasks like Pick or Stow involves a sequence of complex events like moving robot to bin view position, detecting target objects in bin, finding their pose, planning path for the end-effector to reach the object, grasping the object and moving to the drop location. The outcome of one event decides the next action to be carried out. We use SMACH-ROS[3] for software integration. SMACH is a task-level architecture for rapidly creating complex robot behavior. It is suitable where possible states and state transitions can be described explicitly. It allows the developer to maintain, debug large and complex hierarchical state machines. Two different SMACH states are used for implementing strategies for pick or stow task. The first one, `ServiceState` of SMACH-ROS, is used to make service calls to different ROS services, like reading a JSON file, object detection, pose estimation, motion planning, etc. The second one is the generic SMACH state used for other tasks, for instance, to decide when to check for target object in next bin, or when to look for occluded objects and, handling failures, etc. A snapshot of the complete system along with their underlying modules are shown in Fig. 9: (c) object detection module, (d) pose estimation module (e) motion planning module, and (f) state-machine visualization providing run-time status of the system. A video demonstrating the overall working of the system is made available online [16] for the convenience of interested readers.

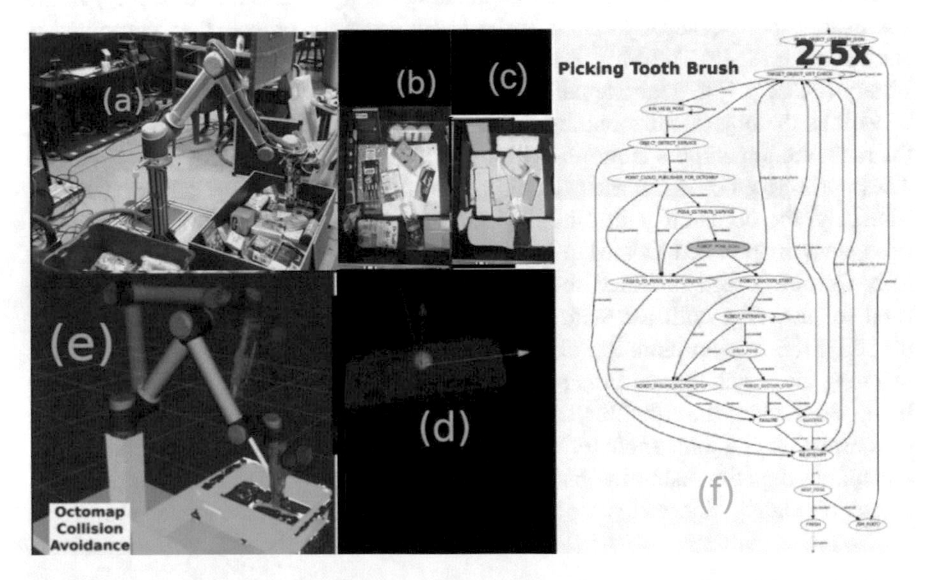

Fig. 9 System integration. (**a**) Actual hardware system, (**b**) view obtained from end-effector camera system, (**c**) semantic segmentation of individual objects as obtained from a deep network based recognition system, (**d**) object pose estimation obtained from 3D depth point cloud, and (**e**) Motion planning required for picking the object while avoiding collision with obstacles

[3]http://wiki.ros.org/smach.

Table 2 Computation time for various modules of the robotic pick and place system

S. No.	Component	Description	Time (s)
1	Reading JSON file	For ID extraction	00.01
2	Motion 1	Home position to Bin View Position	3.5
3	Object recognition	using trained RCNN model	2.32
4	Motion 2	Pre-grasp motion	9.6
5	Motion 3	Post-grasp motion	4.97
6	Motion 4	Motion from Tote drop to home position	3.41
	Total loop time for each object		23.81
7	Rack detection		2.1
8	Calibration		13.1

8 System Performance

The computation time for different modules of the robotic pick and place system is provided in Table 2. As one can see the majority of time is spent in image processing as well as in executing robot motions. Our loop time for picking each object is about 24 s which leads to a pick rate of approximately 2.5 objects per minute. The rack detection and system calibration is carried out only once during the whole operation and does not contribute towards the loop time.

9 Conclusions

In this chapter, we provide an overview of the system developed by IITK-TCS team for ARC 2017 challenge held in Nagoya, Japan. The hardware system comprised of a 6 DOF UR10 robot manipulator with an eye-in-hand 2D/3D vision system and a suction end-effector for picking objects. Some of the novel contributions made while developing this system are as follows. (1) An automated annotation system was developed that could generate large number of templates and train a deep network to recognize and segment 40 objects in a clutter. The training and data generation was completed within the specified 45 min prior to the start of the challenge. (2) A novel hybrid gripper was designed and developed that combined suction capability with a two-finger retractable gripper. (3) A novel geometry based grasping algorithm was developed that allowed us to compute graspable affordances for objects in extreme clutter. The grasping algorithm solves the grasp pose detection problem independent of the object recognition and makes use of only depth point cloud and gripper geometry to solve the grasping problem. The whole system was integrated using SMACH-ROS architecture which allowed us to incorporate complex decision-making capabilities into the system. Overall, we could achieve a pick rate of about 2.5 objects per minute. The team secured fifth position in stow task, third position in pick task, and fourth position in the final round.

Acknowledgements We acknowledge the contribution of many members who worked for the IITK-TCS team. Some of the foremost members in this list include Ashish Kumar (IITK), Ravi Prakash (IITK), Siddharth (IITK), Mohit (IITK), Sharath Jotawar (TCS), Manish Soni (TCS), Prasun Pallav (TCS), Chandan K Singh (TCS), Venkat Raju (TCS) and Rajesh Sinha (TCS).

References

1. Krizhevsky, A., Sutskever, I., Hinton, G.E.: ImageNet classification with deep convolutional neural networks. In: Advances in Neural Information Processing Systems, pp. 1097–1105 (2012)
2. Ren, S., He, K., Girshick, R., Sun, J.: Faster R-CNN: towards real-time object detection with region proposal networks. In: Advances in Neural Information Processing Systems, pp. 91–99 (2015)
3. Liu, W., Anguelov, D., Erhan, D., Szegedy, C., Reed, S., Fu, C.-Y., Berg, A.C.: SSD: Single shot multibox detector. In: European Conference on Computer Vision, pp. 21–37. Springer, Cham (2016)
4. He, K., Gkioxari, G., Dollár, P., Girshick, R.: Mask R-CNN. In: 2017 IEEE International Conference on Computer Vision (ICCV), pp. 2980–2988. IEEE, Piscataway (2017)
5. Zhao, H., Shi, J., Qi, X., Wang, X., Jia, J.: Pyramid scene parsing network. In: IEEE Conference on Computer Vision and Pattern Recognition (CVPR), pp. 2881–2890 (2017)
6. Singh, C.K., Majumder, A., Kumar, S., Behera, L.: Deep network based automatic annotation for warehouse automation. In: International Joint Conference on Neural Networks (IJCNN), Rio de Janeiro, Brazil, July. IEEE, Piscataway (2018)
7. Mahajan, K., Nanduri, H., Majumder, A., Kumar, S.: A deep framework for automatic annotation with application to retail warehouses. In: 29th British Machine Vision Conference (BMVC), Newcastle, UK, September (2018)
8. Raju, V.: Hybrid gripper version 3.0 (2017). https://www.youtube.com/watch?v=wOcMFIXzeuU
9. Gualtieri, M., ten Pas, A., Saenko, K., Platt, R.: High precision grasp pose detection in dense clutter. CoRR, abs/1603.01564 (2016)
10. Kundu, O., Kumar, S.: A novel geometry-based algorithm for robust grasping in extreme clutter environment. Preprint. arXiv: 1807.10548 (2018)
11. Ensenso: 3D depth cameras from ids imaging (2018). https://en.ids-imaging.com/ensenso-n35.html
12. Itseez: Open source computer vision library (2015). https://github.com/itseez/opencv
13. Pan, J., Chitta, S., Manocha, D.: FCL: A general purpose library for collision and proximity queries. In: 2012 IEEE International Conference on Robotics and Automation (ICRA), pp. 3859–3866. IEEE, Piscataway (2012)
14. Wurm, K.M., Hornung, A., Bennewitz, M., Stachniss, C., Burgard, W.: OctoMap: a probabilistic, flexible, and compact 3D map representation for robotic systems. In: Proceedings of the ICRA 2010 Workshop on Best Practice in 3D Perception and Modeling for Mobile Manipulation, vol. 2 (2010)
15. Beeson, P., Ames, B.: TRAC-IK: an open-source library for improved solving of generic inverse kinematics. In: 2015 IEEE-RAS 15th International Conference on Humanoid Robots (Humanoids), pp. 928–935. IEEE, Piscataway (2015)
16. Kumar, A., Soni, M., Jotawar, S., Pallav, P.: SMACH-ROS implementation of IITK-TCS ARC system (2017). https://www.youtube.com/watch?v=u3rOB1G8PXU&t=145s

Designing *Cartman*: A Cartesian Manipulator for the Amazon Robotics Challenge 2017

Jürgen Leitner, Douglas Morrison, Anton Milan, Norton Kelly-Boxall, Matthew McTaggart, Adam W. Tow, and Peter Corke

1 Introduction

Efficient warehouse pick and place is an increasingly important industry problem. Picking specific items out of a box of objects is also one of the canonical problems in robotics. Amazon Robotics has over the years created systems that enable the automated and efficient movement of items in a warehouse. In 2015 they initiated a competition focusing on the problem of how to perform picking and packing tasks with robotic systems.

The 2017 Amazon Robotics Challenge comprised of two tasks, stow and pick, reflecting warehouse operations for online order fulfilment. These involve transferring items between Amazon's red plastic container, a storage system (previously

---

This work was done by J. Leitner prior to joining LYRO Robotics.
This work was done by the author "A. Milan" prior to joining Amazon.
This work was done by the author "A. W. Tow" prior to joining Dorabot.

J. Leitner (✉)
LYRO Robotics, Brisbane, QLD, Australia
e-mail: juxi@lyro.io

D. Morrison · N. Kelly-Boxall · M. McTaggart · P. Corke
Australian Centre for Robotic Vision (ACRV), Queensland University of Technology (QUT), Brisbane, QLD, Australia

A. Milan
Amazon Research, Berlin, Germany
e-mail: antmila@amazon.com

A. W. Tow
Dorabot, Tingalpa, QLD, Australia
e-mail: adam.tow@dorabot.com

A. Causo et al. (eds.), *Advances on Robotic Item Picking*,
https://doi.org/10.1007/978-3-030-35679-8_11

an Amazon provided shelf), and a selection of standard cardboard shipping boxes. Teams were required to develop a robotic solution and were free to design their own storage system within certain limitations. The robots need to visually detect and identify objects in clutter and then successfully transfer them between locations, demanding a robust integration of manipulator, object recognition, motion planning and robotic grasping.

In this chapter we present an overview of our approach and design that led to winning the 2017 competition finals, held in Nagoya, Japan. The primary differentiating factor of our system is that we use a Cartesian manipulator, nicknamed *Cartman* (Fig. 1). We find it to greatly simplify motion planning and executing in the confines of the box shaped storage system compared to articulated robotic arms. It also enabled us to use a dual-ended end-effector comprising two distinct tools. *Cartman* stowed 14 (out of 16) and picked all 9 items in 27 min, scoring 272 points. We released four tech reports explaining the various sub-systems in more detail [17, 18, 20, 29].

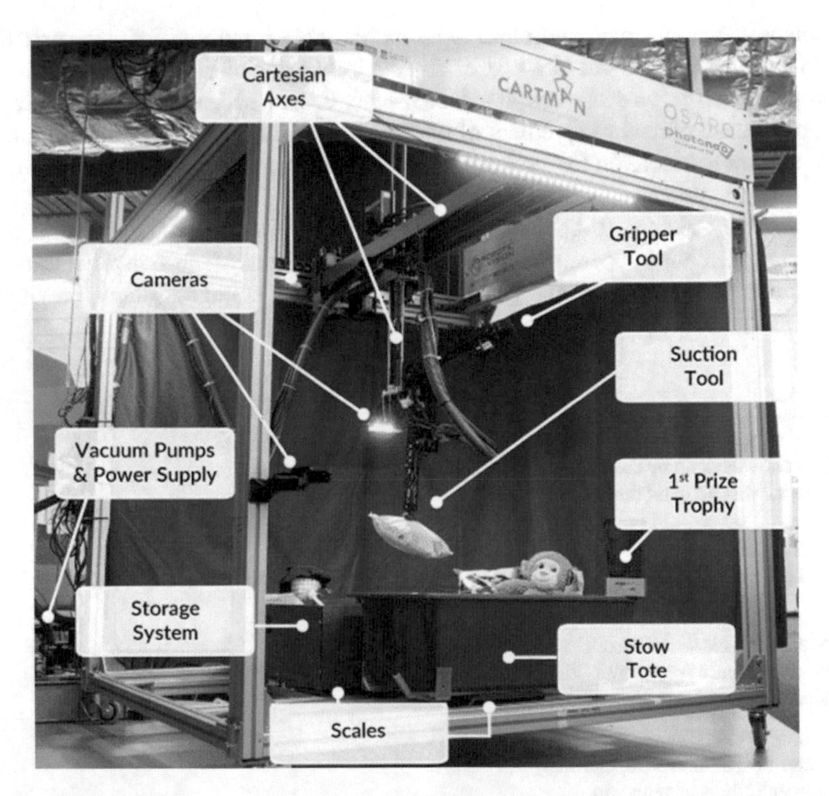

Fig. 1 *Cartman* includes three linear axes to which a wrist is attached. It holds a camera and two end-effector modalities (suction and parallel gripper) that share two revolute axes, and have an extra revolute axis each. To deal with the uncertainty we added a secondary camera, positioned on the frame to take images of picked items with a red backdrop (curtains), and scales underneath the boxes

2 Background

There is a long history of research into grasping and dexterous manipulation [28], and recent progress has largely been driven by technological developments such as stronger, more precise and more readily available collaborative robotic arms [15] and simple compliant and universal grippers [1, 19]. Robust perception is also a key challenge [4]. Over recent years, significant leaps in computer vision were seen thanks to the application of deep learning techniques and large scale datasets [5]. However, the increased performance of computer vision systems has not translated to real-world robotic improvements, highlighting deficiencies in robust perception and hand-eye coordination. Datasets and benchmarks are increasingly exploited in the robotics community to find solutions for such tasks [2, 13] yet still have some shortcomings [12]. The applied nature of competitions makes them one of the greatest drivers of progress—from self-driving cars, to humanoids, to robotic picking.

The 2017 Amazon Robotics Challenge comprised two tasks, stow and pick, analogous to warehouse assignments, which involve transferring items between Amazon's totes (a red plastic container), the team's storage system and a selection of Amazon's standard cardboard shipping boxes. Teams were required to design their own storage system within certain limitations, unlike in previous competitions where standardised shelving units were supplied. Our storage system comprises two red, wooden boxes with open tops.

In the stow task, teams are required to transfer 20 objects from a cluttered pile within a tote into their storage system within 15 min. Points are awarded based on the system's ability to successfully pick items and correctly report their final location, with penalties for dropping or damaging items, or having items protruding from the storage system.

In the pick task, 32 items were placed by hand into the team's storage system. The system was provided with an order specifying 10 items to be placed into 3 cardboard boxes within 15 min. Points were awarded for successfully transferring the ordered items into the correct boxes, with the same penalties applied for mishandled or protruding objects.

The 2017 competition introduced a new finals round, in which the top 8 teams competed. The finals consisted of a combined stow and pick task. Sixteen items were first hand-placed into the storage system by the team, followed by a vigorous rearrangement by the judges. Sixteen more items were provided in a tote and had to be stowed into the storage system by the robot. Then, the system had to perform a pick task of 10 items. The state of the robot and storage system could not be altered between stow and pick.

A major addition to the challenge compared to previous years was that not all items were known to the teams beforehand. The items for each task were provided to teams 45 min before each competition run, and consisted of 50% items selected from a set of 40 previously seen items, and 50% previously unknown items. This change introduced a major complexity for perception, as systems had to be able to handle a large number of new items in a short period of time. This, in particular, made deep learning approaches to object recognition much more difficult.

3 *Cartman*: System Overview and Design

Most pick-and-place robots require at least a manipulator, a gripper, and a perception system. The design of these components and the overall robot system itself, must consider a range of constraints specific to the given application. These constraints will often include: the items the robot must manipulate, the environment in which the robot will operate, required operating speed, cost or return on investment, overall system reliability, grasp success rate, human-safe operation, etc. Herein we describe our final system design (both hardware and software) and how it met the constraints of the competition. We also discuss our design process, of which, we argue, contributed largely to our success in the competition.

Teams almost exclusively competed with articulated robotic arms [4], yet the task of picking items from a storage system is mostly a linear problem, which can easily be broken down into straight line motions. We had previously, in the 2016 challenge, competed with a Baxter robot [13], and encountered difficulties in planning movements of its 7-DoF arm within the limited confines of the shelf. Linear movement of an articulated arm requires control of multiple joints, which if not perfectly executed may result in sections of the arm colliding with the environment.

Our 2017 challenge robot, *Cartman*, in contrast, greatly simplifies the task of motion planning within a storage system due to with its Cartesian design and ability to move linearly along all three axes. We find considerable benefits of using such a design in warehouse pick-and-place tasks:

- *Workspace*: Cartesian manipulators have a compact rectangular workspace, compared to a circular of arms. An advantage particularly for the ARC constraints.
- *Simplicity*: Motion planning with Cartesian robots is simple, fast, and unlikely to fail even in proximity to shelving.
- *Predictability*: The simple, straight-line motions mean that failed planning attempts and erratic behaviour are less likely.
- *Reachability*: Cartesian design results in improved reachability (see Fig. 3) within the workspace and lowered chance of collision compared to arms.

The software developed for *Cartman* is utilising ROS (Robot Operating System) [23], which provides a framework for integrating the various sensors, processing sub-systems, and controllers. As per the regulation of the challenge, it is fully open-source and freely available for download.[1]

[1] https://github.com/warehouse-picking-automation-challenges/team_acrv_2017.

Storage System In 2017 the teams were allowed to design their own "storage system", replacing the previously used Kiva shelf provided by Amazon. The constraints were a total bounding-box volume of 5000 cm^3 (roughly the volume of two Amazon totes) and between 2 and 10 internal compartments. This major difference to the previous competitions opened a new design space for the teams. Our design consists of two red, wooden boxes approximately matching the dimensions of the Amazon totes, and opts for a horizontal (top-down picking) design instead of a vertical shelf-like design, which the large majority of teams concluded to be the more reliable approach. The similarity in colour to the totes allows the same perception system to be used for both the tote and the storage system.

4 Mechanical Design

As mentioned, *Cartman* employs a Cartesian design as depicted in Fig. 1. Articulated manipulators are common for many robotics applications due to their versatility and large workspace with respect to their mechanical footprint. They though have singularities and discontinuities for certain end-effector configurations [3, 9]. There are ways of reducing but not eliminating these drawbacks on motion planning [6, 21]. Using a Cartesian manipulator with a wrist joint to work in a Cartesian workspace eliminates almost all singularities and discontinuities (see Fig. 3 for a comparison of *Cartman* and other robot arms). On the other hand the disadvantage with a Cartesian manipulator is the requirement for a larger mechanical footprint in ratio to the overall workspace of the manipulator. This is due to the fact that the linear motion requires some form of support, usually a rail, along the entire length of the axis. This is less of a problem in warehouses and other conveyor-belt type operations, where the operation space can be completely enclosed by the robot.

4.1 Specifications

The entire manipulator system is mounted on frame of aluminium extrusions and rails for the three axes (Fig. 2). Apart from the three linear axes it consists of a wrist joint controlling roll, pitch, and yaw of a multi-modal end-effector also developed for the challenge. The following specifications were set out before designing the manipulator for the Amazon Robotics Challenge:

- A reachable workspace of 1.2 m $\times$ 1.2 m $\times$ 1.0 m
- A top linear velocity of 1 m/s under load along the three linear axes ($X/Y/Z$).
- A top angular velocity of 1rad/s under load along the angular axes (roll/pitch/yaw).
- A payload capacity of 2 kg.

- Six DoF at the end-effector, given by three linear axes forming the Cartesian gantry and a three-axis wrist.
- Ability to be easily de/constructed to simplify transport of the robot overseas to the event.

4.2 Mechanical Components

The manipulator frame is mounted on a stand of standard aluminium extrusions as seen in Fig. 2. The **frame** consists of laser cut and folded 1.2 mm sheet aluminium for housing the rails, providing a great trade-off between stability and weight. The reduced weight was important as the robot was to be transported overseas for the competition. The sheet aluminium formed the main outer frame housing the X-axis belt system and transmission rod.

The linear **rails** used for the X- and Y-axes are TBR20 and TBR15 precision type profile rail, respectively. Smaller rails were used in the Y-axis to reduce the overall weight of the system. The Y-axis consists of two 10 mm round rails. The downside to using 10 mm rails, however, is that when the Z-axis is extended a

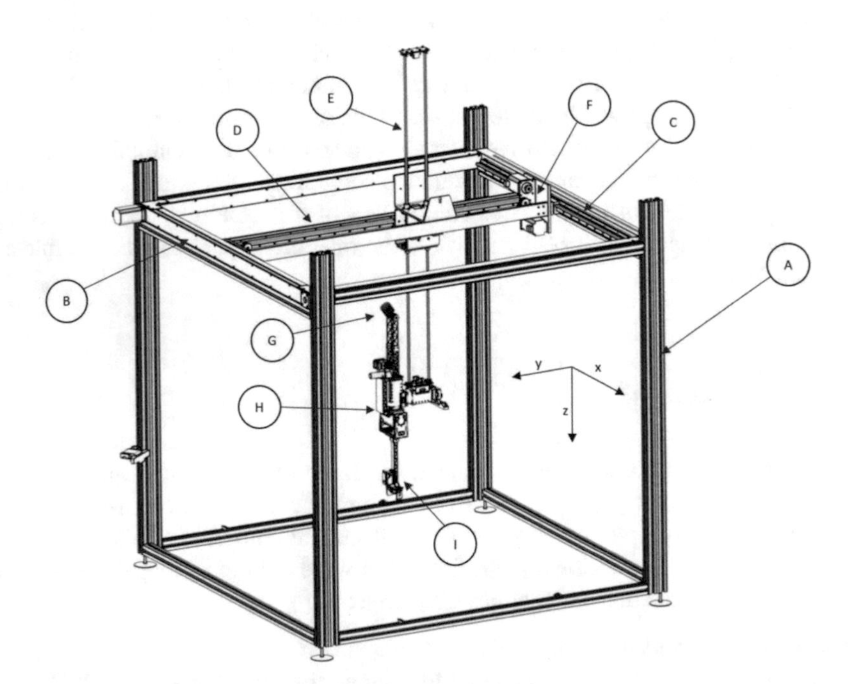

Fig. 2 Isometric view of the entire manipulator. Key components have been labelled and are as follows: (**a**) Aluminium T-slot stand, (**b**) Manipulator Aluminium frame, (**c**) X-axis TBR20 profile rails, (**d**) Y-axis TBR15 profile rails, (**e**) Z-axis 10 mm round rails, (**f**) Y–Z motor-carriage, (**g**) Suction gripper, (**h**) Wrist, (**i**) Parallel plate gripper

pendulum effect is created inducing oscillations at the end-effector due to the rail's deflection. Although deflection and oscillations are present, steady state accuracy is still achieved with ease once settled. We considered this trade-off during the design process. Additionally, the oscillations are minimised by raising the Z-axis when performing large translational movements.

Linear motion is performed by Technic's ClearPath SD-SK-2311S **motors**. They are closed-loop brush-less motors designed to be a drop in replacement for stepper motors, eliminates the need for external encoders. They were chosen due their high performance and ease of use. Three Dynamixel Pro L54-50-500 are controlling the roll, pitch, and yaw axes. These provide the necessary operating torque to hold a 2 kg payload while under acceleration.

To actuate the prismatic joints, a belt and pulley system was used. A single motor drives the X-axis. In order to eliminate a cantilever effect on this, a transmission rod is used to transmit power from one member to another. One common design that has been observed in a lot of simple manipulator designs is each axis motor needs to carry the weight of all distal motors as well as the payload. As a result, more powerful motors are required which increases weight as well as cost. To solve this problem a differential belt system was designed. Rather than using a single motor to drive a single axis, two motors work in tandem to drive two axes (Fig. 3).

4.3 Software and Electrical Components

A single microcontroller is used to control all six axes. We employed a Teensy 3.6 with a breakout board including a logic shifter circuits for each of the ClearPath motor pins. In order to interface with the Dynamixel Pro motors, an RS485 module was added. ROS JointState message type which was processed by the Teensy. The low level firmware functions send commands to both the ClearPath

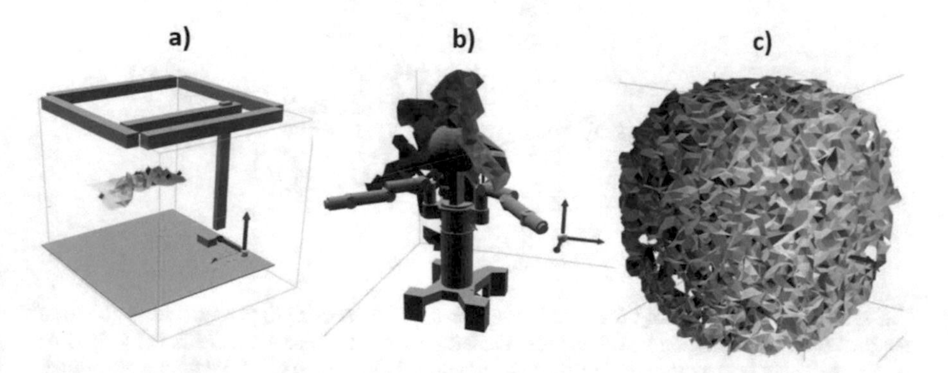

Fig. 3 Discontinuity maps of (**a**) *Cartman*'s end-effector, (**b**) a Baxter's left gripper, and (**c**) a UR-5. If the end-effector passes through these boundaries, joint velocities can accelerate to infinity. *Cartman*'s design limits these discontinuity boundaries, making planning simpler and safer

and Dynamixel Pro motors and also read any feedback that was available from the motors. As the ClearPath motors are a drop-in replacement for stepper motors and as such the open source AccelStepper library [16] is used. It provides the ability to deploy an acceleration profile per motor. The Dynamixel Pro is controlled using a slightly modified version of the OpenCR library [10]. ROS (the Robot Operating System) [24] is handling the higher-level (system) functionality. Desired end-effector poses and robot states are ROS messages published by the microcontroller and used by the MoveIt! package [27], which was the interface to the state-machine.

The complete design is open sourced and available online.[2] For a more in-depth analysis of the design of *Cartman* readers are referred to our tech report [17].

4.4 Multi-Modal End-Effector

The challenge requires teams to pick a very diverse set of items, including rigid, semi-rigid, hinged, deformable, and porous objects. We employ a hybrid end-effector design (Fig. 4) comprising vacuum suction and a parallel plate gripper. Due to the use of a Cartesian system, we do not have to combine suction and gripping into a single tool, leading to a less complex end-effector design (e.g. [7, 25]). We integrate these two distinct tools at the wrist, to be swapped by a single motor rotation. A further advantage, particularly during development, was that this design allows each tool to be developed and tested individually reducing dependencies and downtime.

The grasping system relies on a tight coupling of the physical end-effector and its relevant software components, which are discussed in Sect. 5.5.

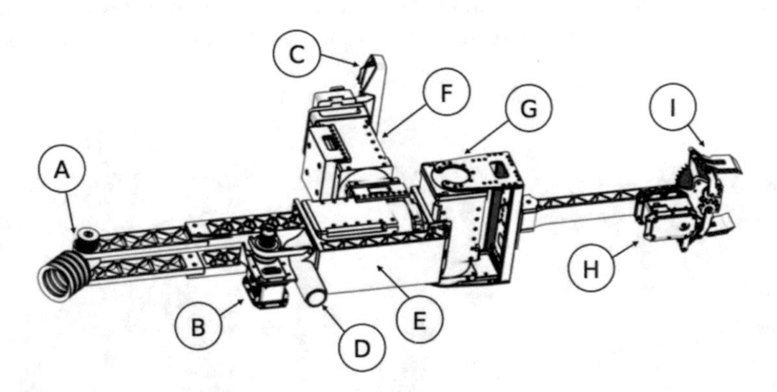

Fig. 4 End-effector Assembly. (**a**) Rotating suction cup. (**b**) Suction gripper pitch servo (drive belt not pictured). (**c**) wrist-mounted RealSense camera. (**d**) suction hose attachment. (**e**) Roll motor. (**f**) Yaw (tool-change) motor. (**g**) Gripper pitch motor. (**h**) Gripper servo. (**i**) Parallel plate gripper

[2]http://Juxi.net/projects/AmazonRoboticsChallenge/.

Vacuum Suction Tool We opted for vacuum suction as the primary grasping mechanism, based on previous year's experience [4]. It lends itself to simpler grasp planning, as only a single, rotation-invariant, grasp point is required to be detected. It has in the past shown an outstanding ability to grasp a large range of items.

The suction system consists of a 40 mm diameter multi-bellow suction cup. It can be rotated in a full semi-circle at its base, allowing it to attach to vertical surfaces inside cluttered environments. Various approaches were tested before settling on a belt-driven (by a Dynamixel RX-10 servo) implementation, which had the best trade-off in terms of overall size, rotation, and reachability. The suction gripper is our default tool, with the parallel gripper being specified for porous or deformable objects which create hardly any vacuum seal, e.g., cotton gloves or marbles.

To get a working prototype for full end-to-end system testing, the initial end-effector design was a simple, rigid suction tool with no articulation (Fig. 5a), based on our previous design [13]. Even using just the initial design, our first *Cartman* prototype was able to attach to 80% of the objects provided. The main limitations identified from the first test were: firstly, to incorporate gripping as a secondary grasping modality is required, and secondly, improving robustness, particularly in more challenging, cluttered environments. We observed drastically reduced grasp success when unable to approach the grasp at a perpendicular angle, especially encountered with objects leaning against the side of the storage system.

To overcome this articulation of the suction tool was added. At first the entire tool arm was moved in an attempt to keep the physical footprint to a minimum (Fig. 5c). However, this still required a clear approach trajectory. An extra degree of freedom was therefore added to the suction cup. In the final design an extra motor is added, as well as the "arm" extended to 240 mm. The belt-driven design pivots the suction cup closer to the end-point (Figs. 5d), keeping the footprint to a minimum. This

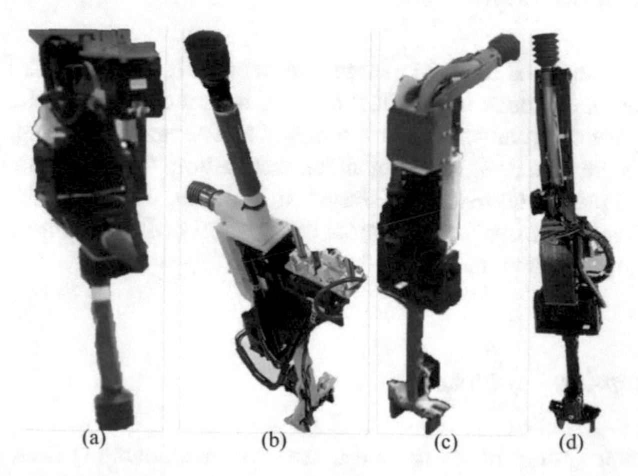

Fig. 5 Four major stages of the end-effector design, as discussed in Sect. 4.4. (**a**) Static suction tool. (**b**) Addition of gripper. (**c**) Extra degree of freedom on suction tool. (**d**) Final design

design allows for 6 degrees of control at the suction end-point, reducing the issues observed in earlier versions.

Parallel Gripper The subset of known Amazon items that could not be picked with our suction system were predominantly those which were porous, or too thin for a suction cup to be attached. We opted to use a parallel gripper as the second grasping modality. A survey of commercially available and open-source parallel grippers did not yield a promising solution for Amazon's challenge tasks, with the options either too large or could not easily be modified.

We opted therefore, to design a custom gripper purpose-built for the challenge with integration to our Cartesian manipulator in mind. By limiting our design parameters to the set of objects which could not be grasped via suction, we created a highly customised solution without any unnecessary overhead, limiting the design space for our gripper. The final design of the parallel jaw gripper uses a single actuating servo and has a stroke of 70 mm, and a maximum grasp width of 60 mm. A 190 mm extension arm connects the wrist to the gripper to ensure it can reach within the Amazon tote.

Our gripper plates feature slightly angled tips (after trialling various designs) to scoop thin objects, such as the scissors, and create a more powerful pinch at the tip for picking deformable objects, such as the sponges or gloves. In addition, high-friction tape add a small amount of compliance to the gripper, specifically helping with rigid objects, while increasing the overall grasp success rate.

For a more in-depth analysis of the end-effector design readers are referred to our tech report [29].

5 Perception System

Manipulating individual items in a large warehouse is a complex task for robots. It is challenging both from a mechanical but also from a computer vision viewpoint. Even though the environment can be controlled to a certain degree, e.g., the storage system can be custom designed to facilitate recognition, the sheer number of items to be handled poses non-trivial challenges. In addition, the items are often placed in narrow bins to save space, thus partial or even full occlusion must be addressed from both the perception and manipulation side.

5.1 Perception Pipeline

The perception system needs to deliver two key functions: (1) detection, identification, and segmentation of the correct item, and (2) the generation of a set of possible grasp points for the object (Fig. 6). The two tasks of the challenge required to detect a variable number of items in the bins, with lighting and placement creating

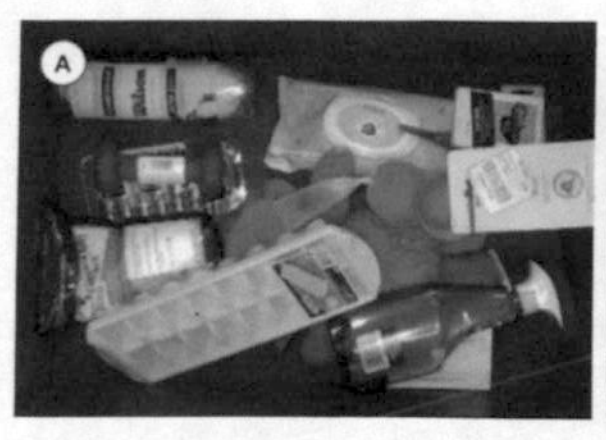

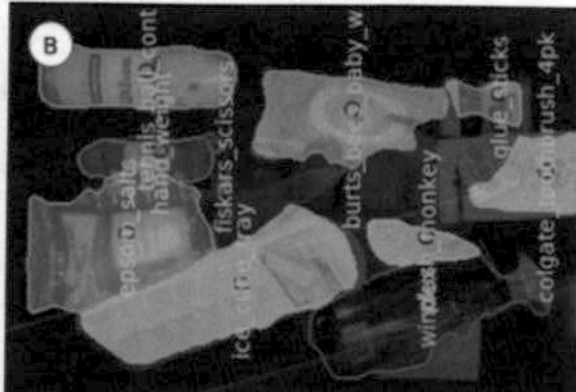

Fig. 6 The perception pipeline, showing (**a**) RGB image of items in our storage system, (**b**) output of semantic segmentation, and (**c**) segmented point cloud for the bottle, with grasp points calculated (higher ranked grasp points shown as longer and green and the best candidate in blue). Even though the semi-transparent bottle results in inaccurate depth information, our system still finds good grasp points on the centre of the bottle, with points on the neck and head being weighted lower

appearance changes throughout the competition. The biggest change, however, to previous edition was in the replacement of 50% of the training objects by new ones. These new items were presented to the participants only 45 min prior to the start of each task. These conditions require a perception system that is very robust and not fully over-fitted to the training set, yet also models that can be quickly adapted to new categories.

In our perception pipeline we perform the two mentioned key functions sequentially. We first perform an object detection, by using a fully supervised semantic segmentation solution based on deep neural networks, followed by running a grasp synthesis technique on the extracted segment. Second-placed Team NimbRo uses a similar perception pipeline, comprising a RefineNet-based architecture which is fine-tuned on synthetic cluttered scenes containing captured images of the unseen items before each competition run [25].

5.2 *Perception Hardware*

An Intel RealSense SR300 RGB-D camera is employed on the wrist as our main sensor. This allows to move the camera actively in the workspace, a feature we exploited by implementing a multi-viewpoint active perception approach described in Sect. 5.4.1. While the camera is light, has a small form factor and provides depth images, a drawback is the infrared projection used to determine each pixel's distance from the sensor. It is unable to produce accurate depth information on black or reflective items. We address this issue with the introduction of alternative grasp synthesis techniques for these items (Sect. 5.5). Furthermore, a second RealSense camera on the robot's frame allows for additional classification of a picked item if required.

In addition, scales are placed underneath the boxes and storage system to measure the weight of the items. These added additional functionalities to perform object identification and error detection. In terms of object classification they provide an addition parameter to verify the correct item is grasped, that it was not dropped, and to detect when the end-effector has come into contact with an item.

A pressure switch is also included on the vacuum line to detect when a suction seal is made, and to detect dropped items.

5.3 Semantic Segmentation

To perform object identification and detection a state-of-the art semantic segmentation algorithm was implemented. It is common practice to fine-tune existing neural network models to specific tasks [8, 22, 26]. This is rather simple in cases where all categories are known and defined beforehand, it is not such a common strategy for tasks where the amount of available training data is very limited, as in the case of the Amazon Robotics Challenge.

After testing various approaches we settled on a deep neural network architecture, RefineNet [14] is able to adapt to new categories using only very few training examples. We use *RefineNet* for pixel-wise classification, and a custom vision-based (as opposed to model-based) grasp selection approach. We argue this to be a more robust and scalable solution in the context of picking in cluttered warehouse environments than those based on fitting models or 3D shape primitives, due to issues with semi-rigid, deformable, or occluded items. The semantic segmentation provides the link between a raw image of the scene and grasp synthesis for a specific object.

5.3.1 Fast Data Collection and Quick Item Learning

Due to the time-critical nature of learning the unseen items during the competition, we developed a semi-automated data collection procedure which allows us to collect images of each unseen item in 7 unique poses, create a labelled dataset and begin fine-tuning of the network within approximately 7 min. Our procedure is as follows:

- Position the wrist-mounted camera above the Amazon tote in the same pose as it would be in during a task.
- From an ordered list, place one unseen item in each half of the tote.
- Maintaining left/right positioning within the tote, change the orientation and position of each item 7 times and capture an RGB image for each pose.
- Using a pre-trained background model, automatically segment each item to generated a labelled dataset. Each automatically generated training image is manually checked and corrected where necessary.
- The new dataset is automatically merged with the existing dataset of known items.
- The RefineNet network is fine-tuned on the combined dataset until shortly before the beginning of the official run approximately 30–35 min. later.

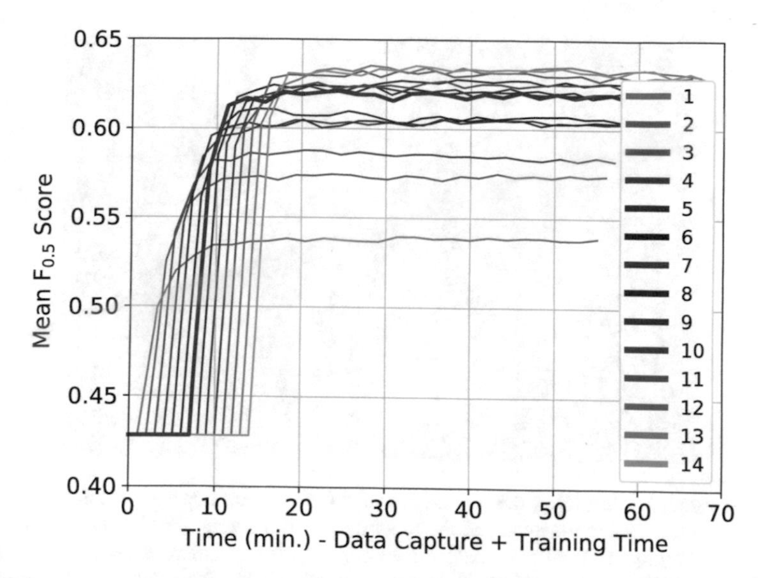

Fig. 7 Mean $F_{0.5}$ score of fine-tuned RefineNet when trained on varying numbers of images for each unseen item. The time includes data capture and training time. For the competition, we used 7 images per unseen item, trained for 13 epochs

Our data collection procedure is a trade-off between time spent on capturing data and available training time. Figure 7 shows the relative network performance for different splits between number of images captured per unseen item and training time. The performance scores are calculated on a set of 67 images representative of those encountered in competition conditions, with a mixture of 1–20 seen and unseen items in cluttered scenes. For the competition, we opted to use 7 images per unseen item, which allows enough time to repeat the data capture procedure if required.

The selection of the metric to be optimised is quite important. We use the $F_{0.5}$ metric to evaluate the performance of our perception system as it penalises false positives more than false negatives, unlike other commonly used metrics such as IOU or F_1 which penalise false positives and false negatives equally. We argue that the $F_{0.5}$ metric is more applicable for the application of robotic grasping as false positives (predicted labels outside of their item/on other items) are more likely to result in failed grasps than false negatives (labels missing parts of the item) (Fig. 8).

As a compromise to cluttered scenes, we opted to capture the images of each new item without clutter, but with as many other commonalities to the final environment as possible. To achieve this, each item was placed in the actual tote or storage container with the camera mounted on the robot's wrist at the same height above the scene as during a run. Each item was manually cycled through a number of positions and orientations to capture some of the variations the network would need to handle.

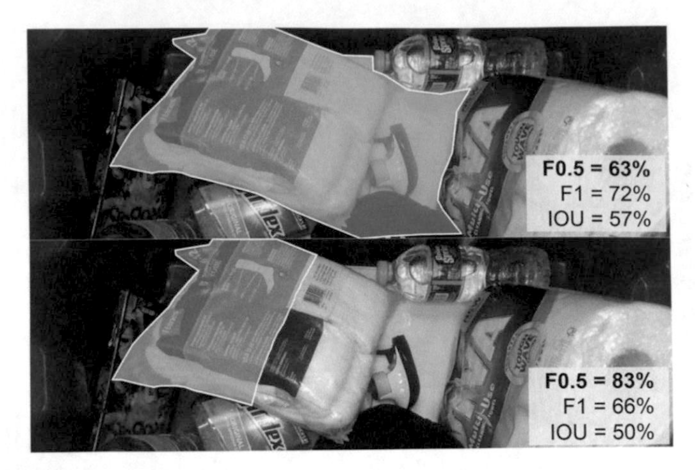

Fig. 8 An example illustrating the importance of different measures for grasping applications. Top: The object is undersegmented and the robot may pick a wrong item if a pick is executed in the segmented region. Bottom: only half of the entire object is segmented correctly. Yet $F1$ and IoU scores are very similar scores as in the above example. The $F_{0.5}$ score is higher than in the above example. We argue the $F_{0.5}$ is therefore more informative and more suitable for creating correct grasp points and successfully manipulate the objects. We argue that the $F_{0.5}$ measure is more informative. Note that precision would also be indicative of success in this example, but should not be used in isolation because it loses information about recall

To speed up the annotation of these images, we employ the same RefineNet architecture as outlined above, but trained on only two classes for binary foreground/background segmentation. After each image is captured, we perform a foreground and background separation. We parallelise this approach by placing two items in the tote at any time. Labels are automatically assigned based on the assumption that the items in the scene match those read out by a human operator with an item list.

During the data capture process, another human operator visually verifies each segment and class label, while manually correcting any flawed samples before adding them to the training set. After a few practice runs, a team of four members are able to capture 7 images of 16 items in approximately 4 min. Three more minutes are required to finalise the manual check and correction procedure.

An in-depth analysis of our RefineNet quick-training approach and a comparison with an alternative deep metric learning approach are provided in [18].

5.3.2 Implementation and Training Details

Training is performed in two steps. Our base *RefineNet* model was initialised with pre-trained ResNet-101 ImageNet weights and initially trained for 200 epochs on a labelled dataset of approximately 200 images of cluttered scenes containing the 40 known items in the Amazon tote or our storage system and an additional background class. Note that the final softmax layer contains 16 (or 10 for the stow task) placeholder entries for unseen categories. Upon collecting the data as described above, we fine-tune the model using *all* available training data for 20 epochs within the available time frame using four NVIDIA GTX 1080Ti GPUs for training. Batch size 1 and learning rate $1e^{-4}$ is used for the initial fine-tuning stage, batch size 32 and learning rate $1e^{-5}$ is used for the final fine-tuning stage. It is important to note that we also exploit the available information about the presence or absence of items in the scene. The final prediction is taken as argmax not over all 57 classes, but only over the set of categories that are assumed to be present.

For a more in-depth analysis and comparison of the object detection system readers are referred to our tech report [18].

5.4 Active Perception

For the stow task, where all items have to be picked, we prefer picking non-occluded items at the top of the tote, which counteracts the tendency for our perception network to exhibit lower precision in more cluttered scenes. However, for the pick task, where only a subset of items have to be picked, it is likely that not all wanted items are easily visible within the storage system. In this task, we use two active perception methods to increase the chances of finding all of the wanted items.

5.4.1 Multi-View

For each storage compartment, there are three possible camera poses; one top view capturing the entire storage compartment, and two close-up views covering half of the storage compartment each. If no wanted objects are visible from the top view, the two close-up views are used, leveraging the adjusted camera viewpoint to increase the chances of finding any partially obscured items, and reducing the effective level of clutter thereby improving the performance of our perception system on the scene.

5.4.2 Item Reclassification

To be sure of a grasped object's identity, our system requires consensus from two sensors. The first is by the primary visual classification. Secondly, when a grasped item is lifted, the weight different measured by the scales is used to confirm the object's identity. If the measured weight does not match, one of the two reclassification methods is used. If the measured weight matches only one item in the source container, then the item is immediately reclassified based on weight and the task continues uninterrupted. Alternatively, if there are multiple item candidates based on weight, the item is held in view of the side-mounted camera to perform a second visual classification of the item. If the item is successfully classified as one of the candidates, it is reclassified and the task continues. If no suitable classification is given, the item is replaced and the next attempt begins.

5.5 *Grasp Synthesis*

The function of the grasp synthesis sub-system is to provide a final pose at which the robot should position its end-effector to successfully grasp the desired item. Our grasp synthesis assumes that the desired object is segmented and will rank possible grasp points on the point cloud segment (Fig. 6) using a hierarchy of approaches. For our vision-based grasp synthesis system, the material properties of an item are of particular importance as they affect the quality of the depth information provided by our camera. To handle a variety of cases, three different grasp synthesis strategies were developed: *surface-normals*, *centroid*, and *RGB-centroid*. If one method fails to generate valid grasp points for an item, the next method is automatically used at the expense of precision. For items where the visual information is known to be unreliable, item meta-data can be provided to weigh specific methods more.

We again use vacuum suction as the primary mode to pick an item, as such we base our grasping pipeline on previous work [11]. The heuristic based ranking is taking into account geometric and visual information, such as distance from edges and the local surface curvature. Some additional end-effector specifics are taken into account as well during the ranking process, e.g., angle to vertical and distance to the workspace edges. This design helps to ensure sensible suction points are ranked high and can be reached with our robot configuration.

If the quality of the point cloud is not allowing for valid grasp points to be detected, as is common with reflective and transparent objects, we approximate the centroid of the available depth points and select it as the grasp point. This method relies on the compliance of the end-effector hardware and the design of both grasping mechanisms which were designed specifically to handle the possible range of objects in the most robust way.

In the worst case, if no valid points are detected in the point cloud, which is most common with black objects, we approximate the centre of the object using its position in the RGB image and estimate a grasp point using the camera's intrinsic parameters. This relies on the design of the end-effector to handle uncertainty in the grasp pose, as well as the mentioned extra sensing modalities of suction detection and weight sensors for collisions.

To grasp an object with the parallel gripper a similar approach including most of the same heuristics was developed. Most importantly, the extra information about the orientation of the object is taken into account to place the gripper in the correct orientation. As a result, the parallel-gripper grasping pipeline requires only one extra processing step compared to the suction system, a simple principle component analysis of the item's RGB segment and align the gripper accordingly.

One aspect of the grasping system which sets *Cartman* apart from other participants is the ability to execute multiple grasp attempts sequentially. For suction grasps, the 3 best, spatially diverse grasp points are stored, and if suction fails on one the next is automatically tried without having to perform another visual processing step, increasing the chances of succeeding in the grasp and increasing overall system speed.

6 Design Philosophy

Our design was driven by the constraints set by the challenge, such as the limits of the workspace, objects to be handled, and the overall task description. Our approach was to test the whole system in challenge-analogue conditions to validate the progress of the overall system, not sub-systems. This integrated design approach, as we are defining it, is one that favours testing of the entire system end-to-end, using the performance of the system as a whole to drive sub-system design. This is contrary to the more commonly used modular system's design approach which favours the design and testing of sub-systems in isolation, which are then, much like a puzzle, fitted together to create the wanted system.

6.1 End-to-End Testing

The frequent end-to-end system testing and the achieved mock scores allowed us to identify the bottlenecks and focus design efforts towards areas that would lead to the largest gain in competition performance. The holistic view of the system enabled design updates to include both software and hardware, as opposed to focusing on the performance of individual sub-systems.

To facilitate rapid hardware testing and integration, rapid prototyping and manufacturing techniques, such as 3D printing and motors with modular mounting options, were employed. This allows for cheap and flexible exploration of new designs as well as the ability to easily produce spare parts in case of failure.

For example, our hierarchical grasp detection algorithm was designed to work with varying levels of visual information. This fully custom design is the result of our integrated design methodology, in which both hardware and software were developed in parallel, with an emphasis on end-to-end system testing throughout the design process, leading to an approach that covered most use cases robustly.

6.2 Modularity

While focus was put on the whole system, we still designed our software and hardware with modularity in mind, making it possible for sub-systems to be developed independently, and easily integrated into the system without requiring changes to higher-level systems. In the case of software, sub-systems largely conform to pre-defined ROS message types throughout development, an example being the perception system which saw major iterations before a final solution was found. Similarly, changes to the manipulator or Cartesian gantry system can be made and easily integrated into the robot's model, leaving higher-level logic unaffected.

6.3 Rapid Iteration

Iterative design is core to the development of our system. Tasks were broken into weekly sprints, and individual design teams were expected to deliver solutions that could be integrated and tested on the system, a process facilitated by our modular design practices. This process results in a higher overall integrated system up-time and allowed the team to focus on testing and evaluating the complete system, and to rapidly improve the design at a system or sub-system level as required.

By designing a fully custom solution, we overcame a major disadvantage reported by teams in previous challenges of being locked into the functionality of off-the-shelf components [4]. Our design comprises many commonly available parts such as the frame and rails for the Cartesian gantry system, simple machined parts, and a plethora of 3D printed parts. As such, many aspects of our design are able to be integrated, tested, and re-designed within hours or days.

7 System Performance

An exhaustive testing of the whole system was performed throughout the development phase as mentioned above. In particular, we were interested in simulating scenarios we would expect to see during the challenge in Japan. This system level testing led to a large focus on robustness and error recovery in developing high-level logic for *Cartman*. Herein we present the results during the Amazon Robotics Challenge finals, as well, as a long-term test of the robot during a robotics event on campus.

7.1 Amazon Robotics Challenge Finals

The finals task of the Amazon Robotics Challenge is chosen as a benchmark for comparison. While all teams received a different set of items, similar object classes were chosen by the Amazon team to be of equal difficulty.

Table 1 compares *Cartman*'s performance in the finals task to the other teams' systems, recorded by matching video footage of the competition with score-sheets. Due to the wide range of strategies used by different teams and the complex environment of the challenge, it is difficult to directly compare the performance of different systems as a whole in the context of the challenge except by points. However, to highlight the strengths and weaknesses of different systems, three metrics for the different aspects of system performance are presented.

Grasp Success Rate The success rate of the system in executing a grasp attempt, regardless of the item or the action performed with it. We counted success as lifting an object, which may be for picking, item classification, or relocating moves.

Table 1 Speed and accuracy of all systems during finals task and our long-term test (ACRV-LT)

Team	Grasp success rate	Avg. time	Error rate	Final score
Applied robotics	50% (3/6)	101 s	0% (0/2)	20
IFL PiRo	78% (18/23)	59 s	50% (7/14)	30
NAIST-Panasonic	49% (21/43)	35 s	33% (5/15)	90
MIT-Princeton	66% (43/65)	25 s	0% (0/15)	115
IITK-TCS	79% (19/24)	40 s	15% (3/20)	170
Nanyang	53% (23/43)	32 s	4% (1/25)	225
NimbRo picking	58% (33/57)	29 s	0% (0/22)	235
ACRV	63% (33/52)	30 s	4% (1/23)	272
ACRV-LT	72% (622/863)	30 s	N/A	N/A

Average Time per Attempt The time from finishing an action with one item to finishing an action with the next, averaged over all actions. It takes into account perception, motion planning, grasp execution, and movement.

Error Rate We define the error rate of the system as the ratio of number of penalties (incorrect item labels, incorrect items picked, dropped/damaged items, etc.) incurred to the total number of items stowed and picked during the round. It is indicative of the overall system accuracy.

While having a high grasp success rate, low execution time and low error rate are all desirable aspects of an autonomous robot, Table 1 shows that no one metric is a good indicator of success. Figure 9 shows the points accumulated by each team throughout the finals task of the competition, including any penalties incurred, and highlights some of the key differentiating aspects of the teams. Performance is most consistent between teams during the stow task, with the top five teams stowing roughly the same number of items at a similar, constant rate. The main separation in points is due to the pick task. For each team, the rate of acquiring points decreases throughout the pick task, as the difficulty of remaining items and the chances of items being occluded increase, causing many teams to abort their attempt early. It is here that we attribute our design approach and our system's overall robustness to our win. During the finals round, our system relied on item reclassification, misclassification detection, failed grasp detection, and ultimately our active perception approach to uncover the final item in the pick task, making us the only team to complete the pick task by picking all available order items.

7.2 Long-Term Testing

To test the overall performance of the system, *Cartman* was run for a full day, performing a continuous finals-style task, where the pick phase was used to replace all items from the storage system into the tote. 17 items were used, consisting

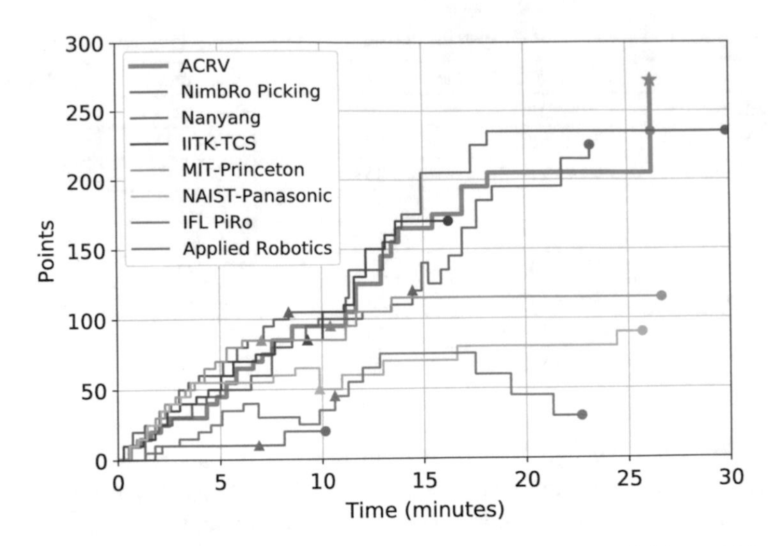

Fig. 9 Points accumulated by each team throughout their finals run, recorded by matching video footage of the competition with score-sheets. Triangles indicate the transition from stow to pick and circles indicate the end of a run. Stars indicate bonus points for completing the pick task with time remaining

of 13 items from the Amazon item set and 4 unseen items from the ACRV picking benchmark set [13]. The 4 unseen items were a soft, stuffed animal `plush_monkey`, a reflective, metallic pet's food bowl `pets_bowl`, a scrubbing brush `utility_brush`, and colourful squeaky pets toys `squeaky_balls`, which were chosen for their similarity to unseen items provided by Amazon during the competition. The 17 items were chosen to provide a range of difficulties as well as cover the spectrum of object classes that were available both physically (rigid, semi-rigid, deformable, and hinged) and visually (opaque, partially transparent, transparent, reflective, and IR-absorbing), ensuring that the full range of *Cartman*'s perception and grasping abilities were tested. Nine of the objects were acquired using suction and 8 by the gripper. Arguably the hardest item in the Amazon set, the mesh cup, was not included as *Cartman* was unable to grasp this item when it was on its side.

In 7.2 h of running time, *Cartman* completed 19 stow and 18 pick tasks, during which 863 grasping attempts were performed, 622 of which were successful (72% success rate, ACRV-LT in Table 1). Throughout the experiment, 10 items were incorrectly classified without automatic correction, requiring manual intervention to allow the system to complete a task and continue with the next. On one occasion the system had to be reset to correct a skipped drive belt.

The overall grasping success rates per item are shown in Fig. 10. Grasp attempt failures were classified as *failed grasp*, where the item was not successfully sucked or gripped, *dropped item*, where the object was successfully sucked or gripped but then dropped before reaching its destination, *weight mismatch*, where an item

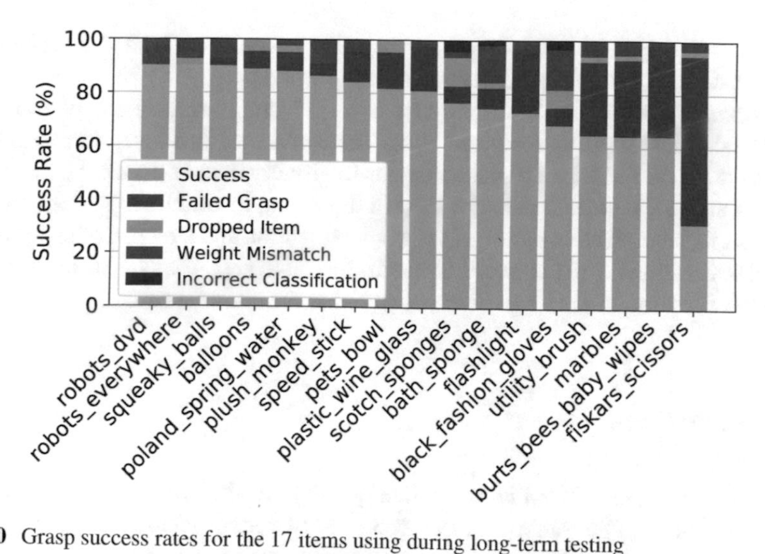

Fig. 10 Grasp success rates for the 17 items using during long-term testing

was grasped but its weight did not match that of the target object, or *incorrect reclassification*, where an object was successfully grasped but was incorrectly reclassified as a different item based on its weight.

The 168 *failed grasp* attempts can be further categorised by their primary cause, either *perception*, where the failure was caused by an incorrect or incomplete segmentation/identification of the object (27.6%), *physical occlusion*, where the object was physically occluded by another, resulting in a failed grasp attempt (10.3%), *unreachable* if the object was in a pose which was physically unobtainable by our gripper such as a small object resting in the corner of the storage system or tote (7.5%), or *grasp pose failure* if the object was correctly identified and physically obtainable and the grasp failed anyway (54.6%). Forty percentage of all failed grasps were on the challenging `fiskars_scissors` item, indicating that our manipulator or grasp point detection could be improved for this item.

7.3 *Error Detection and Recovery*

Like with any autonomous system designed to operate in the real world, things always can and eventually will go wrong. As such, the robot's sensors are monitored throughout the task to detect and automatically correct for failures, such as failed grasps, dropped items, and incorrectly classified or re-classified items. During the long-term testing described in Sect. 7, only 4% (10 out of 241) of failures were not corrected, requiring manual intervention for the robot to finish a task. These failures were caused by grasping an incorrect item that happened to have the same weight as the wanted item, picking two items together and reclassifying by weight as a different item, or dropping the item outside the workspace.

Cartman's actions are heavily dependent on having an accurate internal state of the task, including the locations of all items. As an extra layer of redundancy, towards the end of a task, the system begins to visually double-check the location of items. If an item is consistently detected in the wrong location, the system attempts to correct its internal state based on visual classification and a record of any previous item classifications. This is possible due to the false negative rate of our classifier being lower in uncluttered scenes encountered at the end of a run.

8 Conclusions

Herein we presented *Cartman*, our winning entry for the Amazon Robotics Challenge. We attribute the success of our design to two main factors. In particular, we describe the key components of our robotic vision system, in particular:

- A 6-DoF Cartesian manipulator, featuring independent sucker, and gripper end-effectors.
- A semantic segmentation perception system capable of learning to identify new items with few example images and little training time.
- A multi-level grasp synthesis system capable of working under varying visual conditions.
- A design methodology focused on system-level integration and testing to help optimise competition performance.

Firstly, our method of integrated design, whereby the robotic system was tested and developed as a whole from the beginning. Secondly, redundancy in design. This meant that limitations of individual sub-systems could be overcome by design choices or improvements in other parts of the system.

Acknowledgements This research was supported by the Australian Research Council Centre of Excellence for Robotic Vision (ACRV) (project number CE140100016). The participation of Team ACRV at the ARC was supported by Amazon Robotics LLC.

References

1. Brown, E., Rodenberg, N., Amend, J., et al.: Universal robotic gripper based on the jamming of granular material. Proc. Natl. Acad. Sci. **107**(44), 18809–18814 (2010)
2. Calli, B., Walsman, A., Singh, A., Srinivasa, S., Abbeel, P., Dollar, A.M.: Benchmarking in manipulation research: using the Yale-CMU-Berkeley object and model set. IEEE Robot. Autom. Mag. **22**(3), 36–52 (2015). https://doi.org/10.1109/MRA.2015.2448951
3. Corke, P.I.: Robotics, Vision and Control: Fundamental Algorithms in MATLAB. Springer, Berlin (2011)

4. Correll, N., Bekris, K.E., Berenson, D., et al.: Analysis and observations from the first Amazon picking challenge. IEEE Trans. Autom. Sci. Eng. **15**, 172–188 (2018)
5. Deng, J., Dong, W., Socher, R., Li, L.J., Li, K., Fei-Fei, L.: ImageNet: a large-scale hierarchical image database. In: Computer Vision and Pattern Recognition (CVPR) (2009)
6. Hauser, K.: Continuous pseudoinversion of a multivariate function: application to global redundancy resolution. In: Workshop on the Algorithmic Foundations of Robotics (2017)
7. Hernandez, C., Bharatheesha, M., Ko, W., et al.: Team Delft's robot winner of the Amazon Picking Challenge 2016. In: Robot World Cup, pp. 613–624. Springer, Berlin (2016)
8. Husain, F., Schulz, H., Dellen, B., Torras, C., Behnke, S.: Combining semantic and geometric features for object class segmentation of indoor scenes. IEEE RA-L **2**(1), 49–55 (2016). https://doi.org/10.1109/LRA.2016.2532927
9. Innocenti, C., Parenti-Castelli, V.: Singularity-free evolution from one configuration to another in serial and fully-parallel manipulators. J. Mech. Des. **120**(1), 73–79 (1998). https://doi.org/10.1115/1.2826679
10. Jung, R.L.W.: Robotis OpenCR. https://github.com/ROBOTIS-GIT/OpenCR/tree/master/arduino/opencr_arduino/opencr/libraries/DynamixelSDK
11. Lehnert, C., English, A., Mccool, C., et al.: Autonomous sweet pepper harvesting for protected cropping systems. IEEE Robot. Autom. Lett. **2**(2), 872–879 (2017)
12. Leitner, J., Dansereau, D.G., Shirazi, S., Corke, P.: The need for more dynamic and active datasets. In: CVPR Workshop on the Future of Datasets in Computer Vision. IEEE, Piscataway (2015)
13. Leitner, J., Tow, A.W., Sünderhauf, N., et al.: The ACRV picking benchmark: a robotic shelf picking benchmark to foster reproducible research. In: IEEE International Conference on Robotics and Automation (ICRA) (2017)
14. Lin, G., Milan, A., Shen, C., Reid, I.: Refinenet: multi-path refinement networks for high-resolution semantic segmentation. In: IEEE Conference on Conference on Computer Vision and Pattern Recognition (CVPR) (2017)
15. Matthias, B., Kock, S., Jerregard, H., Källman, M., Lundberg, I.: Safety of collaborative industrial robots: certification possibilities for a collaborative assembly robot concept. In: 2011 IEEE International Symposium on Assembly and Manufacturing (ISAM) (2011)
16. McCauley, M.: AccelStepper: AccelStepper library for Arduino (2010). http://www.airspayce.com/mikem/arduino/AccelStepper/
17. McTaggart, M., Morrison, D., Tow, A.W., et al.: Mechanical design of a Cartesian manipulator for warehouse pick and place. Tech. Rep. ACRV-TR-2017-02, arXiv:1710.00967, Australian Centre for Robotic Vision (2017)
18. Milan, A., Pham, T., Vijay, K., et al.: Semantic segmentation from limited training data. In: Proceedings IEEE International Conference on Robotics and Automation (ICRA) (2018)
19. Monkman, G.J., Hesse, S., Steinmann, R., Schunk, H.: Robot Grippers. Wiley, Hoboken (2007)
20. Morrison, D., Tow, A.W., et al.: Cartman: the low-cost Cartesian Manipulator that won the Amazon Robotics Challenge. Tech. Rep. ACRV-TR-2017-01, Australian Centre for Robotic Vision (2017)
21. Nakamura, Y., Hanafusa, H.: Inverse kinematic solutions with singularity robustness for robot manipulator control. J. Dyn. Syst. Meas. Control. **108**, 163–171 (1986). https://asmedigitalcollection.asme.org/dynamicsystems/article-abstract/108/3/163/425826/Inverse-Kinematic-Solutions-With-Singularity?redirectedFrom=fulltext
22. Pinheiro, P.O., Collobert, R.: From image-level to pixel-level labeling with convolutional networks. In: IEEE Conference on Computer Vision and Pattern Recognition (CVPR) (2015)
23. Quigley, M., Conley, K., Gerkey, B., et al.: ROS: an open-source robot operating system. In: Workshop on Open Source Software of the IEEE International Conference on Robotics and Automation (ICRA) (2009)
24. Quigley, M., Gerkey, B., Conley, K., Faust, J., Foote, T., Leibs, J., Berger, E., Wheeler, R., Ng, A.: ROS: an open-source robot operating system. In: Proceedings of the IEEE International Conference on Robotics and Automation (ICRA), Workshop on Open Source Robotics, Kobe, Japan (2009)

25. Schwarz, M., Lenz, C., Garcia, G.M., et al.: Fast object learning and dual-arm coordination for cluttered stowing, picking, and packing. In: IEEE International Conference on Robotics and Automation (ICRA) (2018)
26. Schwarz, M., Milan, A., Lenz, C., Munoz, A., Periyasamy, A.S., Schreiber, M., Schüller, S., Behnke, S.: NimbRo picking: versatile part handling for warehouse automation. In: IEEE International Conference on Robotics and Automation (ICRA) (2017)
27. Sucan, I.A., Chitta, S.: MoveIt! http://moveit.ros.org/
28. Ulbrich, S., Kappler, D., Asfour, T., et al.: The OpenGRASP benchmarking suite: an environment for the comparative analysis of grasping and dexterous manipulation. In: IEEE/RSJ International Conference on Intelligent Robots and Systems (IROS) (2011)
29. Wade-McCue, S., Kelly-Boxall, N., McTaggart, M., et al.: Design of a multi-modal end-effector and grasping system: how integrated design helped win the Amazon robotics challenge. Tech. Rep. ACRV-TR-2017-03, arXiv:1710.01439, Australian Centre for Robotic Vision (2017)

Index

A. Causo et al. (eds.), *Advances on Robotic Item Picking*,
https://doi.org/10.1007/978-3-030-35679-8

Printed in the United States
by Baker & Taylor Publisher Services